Environmental Division

Held at the 2024 AIChE Annual Meeting

San Diego, California, USA
27-31 October 2024

ISBN: 979-8-3313-1678-5

Printed from e-media with permission by:

Curran Associates, Inc.
57 Morehouse Lane
Red Hook, NY 12571

Some format issues inherent in the e-media version may also appear in this print version.

Printed with permission by Curran Associates, Inc. (2025)

For permission requests, please contact AIChE
at the address below.

AIChE
120 Wall Street, FL 23
New York, NY 10005-4020

Phone: (800) 242-4363
Fax: (203) 775-5177

www.aiche.org

Additional copies of this publication are available from:

Curran Associates, Inc.
57 Morehouse Lane
Red Hook, NY 12571 USA
Phone: 845-758-0400
Fax: 845-758-2633
Email: curran@proceedings.com
Web: www.proceedings.com

TABLE OF CONTENTS

ADVANCED TREATMENT TECHNOLOGIES FOR WATER I

ATMOSPHERIC CHEMISTRY AND PHYSICS: LABORATORY, MODELING, AND FIELD STUDIES

CO2 INDUSTRIAL, ENGINEERING AND RANDD APPROACHES

DESIGN AND ANALYSIS OF CARBON CAPTURE AND NEGATIVE EMISSIONS TECHNOLOGIES - EXPERIMENTAL

DESIGN AND ANALYSIS OF CARBON CAPTURE AND NEGATIVE EMISSIONS TECHNOLOGIES - MODELS

DESIGN AND ANALYSIS OF SUSTAINABLE CARBON CAPTURE AND EMISSIONS CONTROL TECHNOLOGIES

DESIGN AND OPTIMIZATION OF INTEGRATED ENERGY SYSTEMS

EMERGING FRONTIERS IN ENVIRONMENTAL CHEMICAL ENGINEERING

ENVIRONMENTAL DIVISION AWARDS AND HONORS (INVITED TALKS)

ENVIRONMENTAL ISSUES INVOLVING BIOCHAR

FUNDAMENTALS AND APPLICATIONS FOR MUNICIPAL SOLID WASTE TREATMENT AND VALORIZATION

FUNDAMENTALS AND APPLICATIONS FOR WASTE TREATMENT AND VALORIZATION I

FUNDAMENTALS OF ENVIRONMENTAL KINETICS AND FOOD, ENERGY, AND WATER SYSTEMS

POSTER SESSION: ENVIRONMENTAL DIVISION

REMEDIATION OF EMERGING CONTAMINANTS AND LEGACY COMPOUNDS

SUSTAINABILITY FUNDAMENTALS AND METRICS APPLICATIONS

SUSTAINABLE FUEL FROM RENEWABLE RESOURCES

WASTE PLASTIC - RECYCLE, REUSE AND REMEDIATION STRATEGIES I

WATER REUSE AND RECYCLING

MICROPLASTICS CONTAMINATION: RESEARCH, TREATMENT AND MITIGATION

EMERGING TREATMENT TECHNOLOGIES AND CHARACTERIZATION FOR PFAS CONTAMINATION I

ADVANCED TREATMENT TECHNOLOGIES FOR WATER II

EMERGING TREATMENT TECHNOLOGIES AND CHARACTERIZATION FOR PFAS CONTAMINATION II

FUNDAMENTALS AND APPLICATIONS FOR WASTE TREATMENT AND VALORIZATION II

WASTE PLASTIC - RECYCLE, REUSE AND REMEDIATION STRATEGIES II

527a Treatment of Tetracycline Antibiotics in Water and Wastewater By Hay-Derived Engineered Biochar Catalyst

Shengquan Zeng, Eunsung Kan

$FeCl_3$-activated bermudagrass-derived biochar (FA-BC) was first used for tetracycline (TC) adsorption followed by regeneration with Fenton and persulfate (PS) oxidation. FA-BC exhibited an excellent TC adsorption capacity by surface complexation, hydrogen bonding, and π-π EDA interaction, while the TC adsorption process was endothermic and spontaneous. Fenton oxidation only exhibited high regeneration efficiency for TC-spent FA-BC at pH 3, while PS oxidation effectively regenerated the TC-spent FA-BC at pH 3-8. Increasing the temperature from 20 to 50 °C enhanced the regeneration performance with Fenton oxidation, while decreasing with PS oxidation. Increasing the temperature from 20 to 50 °C caused the enhancement of TC diffusion, H_2O_2 reaction rate, and PS reaction rate by the factors of 2.20, 1.58, and 4.07. HO· and SO•– 4 radicals mainly participated in the regeneration of TC-spent FA-BC via Fenton and PS oxidation, respectively. Under the optimal conditions, Fenton and PS oxidation effectively regenerated the TC-spent FA-BC with regeneration efficiencies of 99.41% and 99.24%. Adsorption/Fenton oxidation and adsorption/PS oxidation using FA-BC were also applied to treat TC from a real dairy wastewater, achieving the regeneration efficiencies of 94.99% and 93.36%. Thus, this study indicated great potential for application of FA-BC in TC treatment from water and wastewater via cyclic adsorption/Fenton regeneration and cyclic adsorption/PS regeneration.

527b Sustainable Removal of Pharmaceutical Contaminants from Drinking Water Sources: Integrating Advanced Oxidation and Adsorption Approach

Esther Olagunju, Matthew Alexander

Access to clean and safe drinking water is a fundamental human right (United Nations sustainable goal 6) and is essential for maintaining good health. However, many water sources worldwide are contaminated with emerging contaminants, including pharmaceuticals, personal care products, plasticizers, and flame retardants. Pharmaceuticals, such as antibiotics, hormones, and analgesics, can enter the environment through various pathways, including the discharge of treated wastewater, agricultural runoff, and improper disposal of unused medication. Pharmaceutical emerging contaminants (PECs) have become a significant concern for drinking water sources due to their potentially harmful effects on human health and persistence in the environment for prolonged periods. The grave consequences of pharmaceutical contaminants on human health, such as endocrine disruption, antibiotic resistance, and other chronic illnesses, are well-documented. Thus, it is necessary to develop effective treatment methods for removing these contaminants from all potable water sources. Current treatment methods include microbial degradation, photocatalytic degradation, UV oxidation processes, and ozonation. The excessive cost and energy demand of these methods may not be affordable for small counties and most developing countries. Hence, there is a need for simple and low-energy removal techniques to address the water contamination challenge globally. This study aims to establish an efficient and economically viable approach for eliminating PECs from drinking water sources, achievable through a synergistic application of advanced oxidation and adsorption techniques in a laboratory-scale model. The study centers on eliminating two prevalent PECs, Ibuprofen (IBP) and Ciprofloxacin (CIP), while concurrently exploring the counteractive impact of generic co-contaminants such as humic acid and carbonates on the PEC removal efficiency. Thermo Fisher Vanquish Ultra Performance Liquid Chromatography is used to quantify the pharmaceuticals before and after treatment. Specifically, the study assesses the performance of persulfate, permanganate, and activated carbon. The effect of operating parameters, such as pH, contact time, oxidant and adsorbent dosage, and co-contaminants, on the removal efficiency of PECs is also evaluated. The outcome of this study is insights into developing a sustainable and cost-effective integrated approach for eliminating PECs from drinking water sources. Moreover, the research contributes to understanding co-contaminant effects on PEC removal and provides recommendations for optimizing treatment processes. This study addresses the pressing demand for effective and sustainable techniques to eliminate PECs from potable water sources, thereby ensuring the availability of safe and healthy drinking water for all.

527c Enhancement of Fouling Mitigation through Real-Time Induced Magnetic Vibrations in Spin-Coated Magnetized Membranes

Jasneet Pala, Ryan Tracy, Nima Mahmoodi, Milad Esfahani

Despite the advanced capabilities of membrane technologies, which can produce billions of gallons of water daily, fouling remains a significant challenge. Fouling reduces membrane separation efficiency by compromising permeability and rejection, thereby shortening membrane lifespan and escalating process costs through heightened washing requirements. Chemical modification of membranes, while promising, may lead to drawbacks like reduced stability, selectivity, and flux due to structural changes. In this study, we report a physical and in-situ antifouling approach by investigating the effects of real-time induced magnetic vibrations on the membrane surface to hinder or delay the deposition of foulants. Our previous research demonstrated that attaching iron particles to the backside of the membrane and applying vibrations via an external magnet at 5 Hz, during treatment with 45 ppm of humic acid (HA), led to a significant reduction in fouling. This resulted in only a 10 % decline in flux over 6 hours of filtration, compared to a substantial 33.9 % flux reduction observed over 6 hours with no vibrations for the blank membrane. In continuation with the previous work, we focused on synthesizing magnetic membranes by incorporating iron particles on the backside of the membrane using the spin coating technique to get a uniform distribution of iron nanoparticles across the membrane, unlike mixed matrix membrane (MMM). The primary emphasis of this work is on synthesizing a magnetic membrane utilizing a spin coater with a two-layer coating without altering the pore structure of the membranes and evaluating the vibrational antifouling mechanisms at different frequencies. The performance of the magnetic membranes was investigated under a magnetic field, comparing water flux, rejection, and antifouling properties with blank membranes and mixed-matrix membranes. The iron membrane, when spin-coated, demonstrated an improved flux rate with just a 2% reduction over a 6-hour period at a 10 Hz applied frequency, compared to a 26% reduction observed without vibration.

527d A Zwitterionic Hydrogel-Based Heterogeneous Fenton Catalyst for Water Treatment

Devashish Gokhale, Ian Chen, Wan-Ni Wu, Arthur Monne Gagnaire, Patrick Doyle

Persistent organic pollutants (POPs), including xenoestrogens and PFAS, demand urgent global intervention. Fenton oxidation, catalyzed by iron ions, offers a cost-effective means to degrade POPs. However, numerous challenges like acid dependency, catalyst loss, and toxic waste generation hinder practical application. Efforts to create long-lasting heterogeneous Fenton catalysts, capable of simultaneously eliminating acid requirements, sustaining rapid kinetics, retaining iron efficiently, being compatible with UV light, and being reusable have been unsuccessful. Here, we introduce an innovative heterogeneous zwitterionic hydrogel-based Fenton catalyst, surmounting these challenges in a cost-effective and scalable manner. The hydrogel, hosting individually complexed iron ions in a porous scaffold, exhibits substantial effective surface area and mass transfer-controlled kinetics akin to homogeneous Fenton reactions. Individually complexed ions within the hydrogel can initiate Fenton degradation at neutral pH, eliminating acid additions. Simultaneously, the zwitterionic hydrogel scaffold, chosen for its resistance to Fenton oxidation, forms strong bonds with iron ions, enabling prolonged reuse. Diverging from existing designs, our catalyst proves compatible with UV-Fenton processes and achieves rapid self-regeneration during operation, offering a promising solution for the efficient and scalable degradation of POPs. The study underscores the efficacy of our approach by demonstrating the swift degradation of three significant contaminants - xenoestrogens, pesticides, and PFAS across multiple cycles at trace concentrations.

527e Enhanced Stability and Activity of Series $LaFe_xCu_{1-x}O_3$ Perovskites for Ofloxacin Degradation from Wastewater: DFT Calculation, Mechanisms and Toxicity Evolution

Arvind Kumar, Riona Indhur, Sheena Kumari, Faizal Bux

Introduction

In recent decades, the substantial production of wastewater by the pharmaceutical industry has prompted significant environmental apprehensions, primarily due to its detrimental impact on aquatic ecosystems and biodiversity [1]. A relevant example of pharmaceuticals detected in wastewater is ofloxacin (OFLX) ($C_{18}H_{20}FN_3O_4$), a member of the third generation of fluorinated quinolone antibiotics extensively prescribed for various bacterial infections. Several studies have been reported the levels of OFLX in various water settings, including sewage water (0.89 to 31.7 μg/L), hospital wastewater (35 μg/L), urban wastewater treatment plants (1.8 μg/L), and surface water (0.5 μg/L) [2,3]. Owing to potential toxicity, stable chemical structure, and low biodegradability, OFLX cannot be efficiently treat using conventional treatment methods. In the recent years, sulphate radical-based AOPs (SR-AOPs) have become more popular for degradation of emerging pollutants from aqueous media. Owing to stronger redox potential (2.5–3.1 V), higher selectivity of $SO_4^{\bullet-}$ radical compared to $^{\bullet}OH$ radical, and longer life of (30–40 μs) $SO_4^{\bullet-}$ radical, SR-AOPs have demonstrated the excellent efficacy in degradation of emerging pollutants [4,5]. In order to further enhancing the efficacy of SR-AOPs, a series of perovskites were utilized and activated with peroxymonosulfate (PMS). In the present investigation, OFLX was degraded from aqueous media utilizing a series $LaCu_xFe_{1-x}O_3$ perovskites synthesized by citric sol-gel method and activated with PMS. The synthesized perovskite was characterized by various analytical techniques e.g., FESEM/EDX, TEM, BET, FTIR, XRD, XPS and EIS. Subsequently, the effects of crucial parameters such as PMS dosage (250-1000 g/L), catalyst dosage (250-1000 g/L), solution pH (2-12), and reaction temperature (25-55 °C) were analyzed on OFLX degradation along with PMS consumption in the $LaFe_xCu_{1-x}O_3$/PMS system. To evaluate the bio-toxicity of OFLX and their intermediate products analysis by ECOSAR prediction model. Identification of major reactive oxygen species (ROS) generated in the $LaFe_xCu_{1-x}O_3$/PMS system for ofloxacin degradation along with the economic evaluation of the treatment process. DFT calculations (e.g., Fukui function, electrostatic potential (ESP) mapping, and HOMO-LUMO structure) were evaluated to identification of strong attacking sites on ofloxacin molecule.

Methodology

All the batch experiments for OFLX degradation were conducted in 250 mL three neck glass reactor containing the 100 mL of OFLX solution maintained at 25 °C temperature. The total reflux condenser was fitted on the top of the reactor for preventing the vapour loss. The aqueous solution of OFLX were prepared in Milli-pore water with desired concentrations (5-20 mg/L).The concentration of OFLX was analyzed by HPLC system (Waters India, Ltd.), equipped with C-18 column and UV detector at 295 nm wavelength. The eluent was a binary mixture of methanol and pure Milli-Q water (proportion 25:75 v/v); eluent flow rate was 0.8 mL/min and injection volume of 20 mL (manually injected) in isocratic mode.

Results and conclusions

In order to assess the catalytic activity of synthesized perovskites $LaFe_xCu_{1-x}O_3$ catalysts, some initial experimental trials were conducted for OFLX degradation in aqueous solution. The maximum removal of OFLX was observed 4.65%, 5.12%, 7.82%, and 6.74% with the corresponding catalysts 0.2LFC, 0.4LFC, 0.6LFC, and 0.8LFC, without using any oxidant (PMS) as shown in Fig. 1a. Interestingly, when solely employing 1 mM of the oxidizing agent, maximum removal of OFLX was noticed 5.89% under the pH 7.2 and reaction temperature 25°C. Furthermore, while both catalyst and oxidant doses were introduced simultaneously into aqueous media, all catalysts showed remarkable catalytic efficacy in removing ofloxacin, with removal rates of 26.84%, 28.68%, 34.85%, and 29.78%, under the reaction conditions (e.g., 0.5 g/L catalyst dose, 1 mM PMS dose, pH 7.3, and 25 °C reaction temperature). To better understand the quantitative contribution of reactive oxygen species (ROS) generated in the $LaFe_xCu_{1-x}O_3$/PMS system, a series radical quenching experiments were conducted with various quenchers i.e., tert-butyl alcohol (TBA), L-histidine (LH), and ethanol (EtOH). Combined with intermediates analyzed by GC-MS, DFT calculation and literature survey, the possible transformation pathways of OFLX degradation were proposed systematically. To assess the economic viability of the catalyst, five consecutive cycle experiments were conducted for ofloxacin degradation under the optimum operating condition.

Key conclusions

The present study demonstrates that catalyst 0.6LFC exhibited the excellent degradation performance toward OFLX in the $LaFe_xCu_{1-x}O_3$/PMS system. The quenching experiments reveals that $SO_4^{\bullet-}$ radicals was dominant ROS for OFLX degradation in the $LaFe_xCu_{1-x}O_3$/PMS system. Furthermore, six potential routs of OFLX degradation were proposed based on the DFT calculation study, intermediates identified by GC-MS analysis and existing literature.

References

1. Kumar, B. Prasad, S. Kumari, F. Bux, Mechanistic insight into SO_4^{-}/$^{\bullet}OH$ radical for enhancing stability and activity of $LaMO_3$ perovskite toward detoxification of bulk pharmaceutical wastewater: Stoichiometric efficiency and controlled leaching study, Sep. Purif. Technol. 319 (2023) 123967.
2. Su, J. Li, H. Yuan, B. Wang, Y. Wang, Y. Li, Y. Xing, Visible-light-driven photocatalytic degradation of ofloxacin by g-C3N4/NH2-MIL-88B(Fe) heterostructure: Mechanisms, DFT calculation, degradation pathway and toxicity evolution, Chem. Eng. J. 427 (2022).
3. Patidar, V.C. Srivastava, Mechanistic and kinetic insights of synergistic mineralization of ofloxacin using a sono-photo hybrid process, Chem. Eng. J. 403 (2021) 125736.
4. Liang, K. Gao, A. Zhou, Y. Fang, S. Su, L. Fu, M. Sun, X. Duan, Long-lasting performance of high-flux perovskite membrane for catalytic degradation of organic pollutants, Appl. Catal. B Environ. 327 (2023) 122440.
5. J. Qian, R. Ma, Z. Chen, G. Wang, Y. Zhang, Y. Du, Y. Chen, T. An, B. Ni, Hierarchical Co-Fe layered double hydroxides (LDH)/Ni foam composite as a recyclable peroxymonosulfate activator towards monomethyl hydrazine degradation: Enhanced electron transfer and $^1O_2$ dominated non-radical pathway, Chem. Eng. J. (2023) 143554.

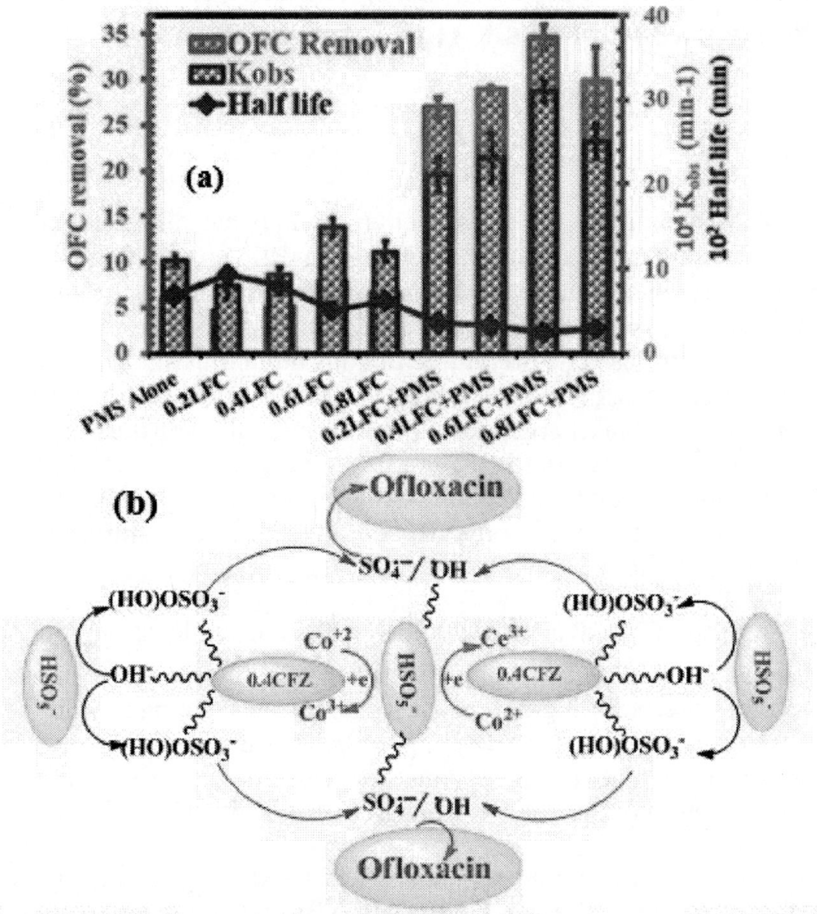

Fig. 1(a) Preliminary experimental study for ofloxacin removal, **(b)** PMS activation mechanism in the $LaFe_xCu_{1-x}O_3$/PMS system.

527f Synthesis of n-Type Metal Doped Molybdenum Diselenide for the Efficient Visible-Light Photocatalytic Degradation of an Antibiotic and Photo-Electrochemical Study

Shubham Raj, Amar Nath Samanta

In recent decades, the discharge of pharmaceuticals into aquatic environments has raised human concern. These pharmaceuticals in waters are derived from hospitals, domestic effluent, and pharmaceutical industries containing excreted or unused medications. Current research findings revealed that the COVID-19 pandemic gave rise to a 102.2% average surge in waste production in both private and government healthcare facilities. The literature contains numerous studies that have documented the levels of CIP in various sources, including as wastewater treatment plants, untreated drinking water, hospital wastewater, lakes, and discharges from pharmaceutical companies. The concentrations vary between 11 and 99 mg/L, 0.032 mg/L, 150 mg/L, 6.5 mg/L, and 31 to 50 mg/L, respectively [1]. The literature review indicates that semiconductor based AOPs can be used to effectively remove pharmaceuticals from wastewater. In order to facilitate the water splitting reaction, it is necessary to use catalysts that are highly active, stable, low cost, and abundant.

Molybdenum diselenide ($MoSe_2$), because of its advantageous flower like morphology and increased conduction band (CB) position, has the potential to be a very effective reducing photocatalyst for the degradation of emerging pollutants and water-splitting process [2]. The intrinsic bandgap value of semiconducting materials can be altered by metal doping, leading to the formation of intermediate energy levels. These energy levels help to separate the photogenerated carriers by increasing the concentration of charge carriers and making it easier for electrons to be trapped [3].

This work involves the production of novel molybdenum diselenide and the doping of zinc metal into it. Using techniques such as FESEM, EDX, XRD, Raman, FTIR, and XPS, extensive investigations will be conducted, which will support the catalyst's efficient production. Using UV-VIS DRS to assess the synthetic sample's optical characteristics, it is possible to identify a band gap that promotes effective visible light absorption. The semiconductor type, charge transfer kinetics, and charge separation and transfer inside the on and off zones will be investigated using the Mott-Schottky plot, Electrochemical Impedance Spectroscopy (EIS) study, and photocurrent investigation, respectively. An appropriate photochemical reactor is used to break down ciprofloxacin (CIP), and the reaction rate constant is measured. Under optimal conditions, the residual concentration of CIP decreased to a point where it is no longer detectable after 45 minutes. Water-splitting will be investigated using chronoamperometry linear sweep voltammetry (LSV), Tafel plot and electrochemically active surface area (ECSA).

References

[1] T. Hayri-Senel, E. Kahraman, S. Sezer, N. Erdol-Aydin, G. Nasun-Saygili, Photocatalytic degradation of ciprofloxacin from water with waste polystyrene and TiO2 composites, Heliyon 10 (2024) e25433. https://doi.org/10.1016/j.heliyon.2024.e25433.

[2] J. Yang, J. Sun, S. Chen, D. Lan, Z. Li, Z. Li, J. Wei, Z. Yu, H. Zhu, S. Wang, Y. Hou, S-scheme 1 T phase MoSe2/AgBr heterojunction toward antibiotic degradation: Photocatalytic mechanism, degradation pathways, and intermediates toxicity evaluation, Sep Purif Technol 290 (2022) 120881. https://doi.org/10.1016/j.seppur.2022.120881.

[3] X. Li, Y. Cheng, Q. Wu, J. Xu, Y. Wang, Synergistic effect of the rearranged sulfur vacancies and sulfur interstitials for 13-fold enhanced photocatalytic H2 production over defective Zn2In2S5 nanosheets, Appl Catal B 240 (2019) 270–276. https://doi.org/10.1016/j.apcatb.2018.09.008

527g Sustainable Degradation of Ciprofloxacin in Water By the Electro-Peroxone Process Via a Graphite Felt Electrode System

Ramya Srinivasan

Over the last few decades, excessive, unregulated usage and reckless disposal of antibiotics have led to the development of antibiotic resistance in bacteria. Additionally, untreated or partially treated pharmaceutical effluents are often discharged into water bodies in an unregulated manner, especially in developing countries. With the increasing prevalence of antibiotic resistance across the globe, previously curable diseases are becoming challenging to handle due to the advent of superbugs. It is imperative to ensure a complete treatment and removal of antibiotics and pharmaceuticals before discharging them into water bodies. Unfortunately, conventional treatment plants are not adept in removing such complex, emerging contaminants. The field of advanced oxidation processes is a rapidly evolving and promising treatment strategy for the abatement of emerging contaminants such as pharmaceutical compounds. Electro-peroxone, a propitious advanced oxidation process, is investigated in this work for the treatment of pharmaceuticals, by considering ciprofloxacin, a model antibiotic drug as the target contaminant. The electro-peroxone system houses electrodes of graphite felt that help form hydrogen peroxide, an oxidant, in situ (see schematic). The viability of the carbon-based graphite felt system in the generation of reactive oxygen species and in ciprofloxacin degradation is investigated in detail. A comparison of the electro-peroxone with two benchmark processes, namely, electrolysis and ozonation, is also carried out. Both electro-peroxone and ozonation are show promise in the removal of the model contaminant and the in situ generation of H2O2 and •OH is estimated. This is a one-of-a-kind study involving graphite felt as both anode and cathode and achieves an in situ generation of H2O2 of 47 mg/L in 120 min and •OH of 140 μM within 60 min of electroperoxone. Besides, the efficacy of the system in contaminant degradation is examined at voltammetric and galvanostatic modes of operation. Ozonation and electro-peroxone processes achieved an efficiency between 97% and complete removal of ciprofloxacin in less than 1 hour. This novel system generates several times higher hydrogen peroxide than the existing graphite electrode system, making it more efficient in radical generation and pollutant abatement (see illustration). This graphite felt-based electro-peroxone system, on further optimization and up-scaling, can be a promising strategy for abating pharmaceutical compounds and effluents.

Reference:

Srinivasan, Ramya. 'Sustainable degradation of ciprofloxacin in water by the electro-peroxone process via a graphite felt electrode system.' *Discover Water* 4.1 (2024): 1-8.

468a Changing Idling Behavior through Dynamic Air Quality and Idle Detection Messaging

Tristalee Mangin, Kerry Kelly, Xiwen Li, Saba Mahmoudi, Rehman Mohammed, Ross Whitaker, Greg Madden, Pierre-Emmanuel Gaillardon, Evan Blanchard, Nathan Page, Ashton Snelgrove

Concentrated vehicle engine idling can cause microenvironments of poor air quality, and areas with high idling, such as schools or hospitals, are frequented by individuals at increased risk for negative impacts from poor air quality. Anti-idling signage and education have been used to influence driver idling behavior. However, these strategies have had limited success. In this study, we try a different approach – dynamic feedback about air quality and idling status. A similar approach, dynamic feedback on driver speed, has become commonplace and has effectively reduced speeding and traffic accidents in the vicinity of the speed display. Our study aims to evaluate how dynamic messaging about idling behavior and air quality affects driver idling choices. Our study entailed developing a system that integrates networked low-cost air quality sensor measurements and idling vehicle detection based on video and audio inputs to provide dynamic feedback to drivers. The dynamic feedback system comprises seven air quality sensor nodes, one video camera, six microphones, a central server, and two outdoor displays to provide feedback to the drivers regarding idling status and local air quality. The low-cost air quality sensor nodes measured carbon dioxide (CO2), carbon monoxide (CO), nitric oxide (NO), nitrogen dioxide (NO2), total volatile organic compounds (TVOC), and particulate matter (PM2.5). The video camera and the six microphones provide information about vehicle motion and engine noise. The team developed a machine-learning technique to identify idling vehicles from the audio and video inputs. The server then integrated the air-quality sensor measurements and idling status (from the audio and video inputs) to provide real-time feedback about vehicle idling and air quality in messages sent to the outdoor displays. We deployed our dynamic feedback system at a hospital's drop-off zone, where numerous vehicles frequently idle. This deployment lasted 15 days. Five days showed a static control message on the outdoor displays; five days showed a machine learning idle status-based dynamic message; and five days showed an air quality metric-based dynamic message. The machine learning algorithm performed well over the 15-day deployment, with an average idling label precision of 78.3% and an average non-idling precision of 96.6% compared to onsite notes. The dynamic message deployment days resulted a 40% decrease in CO2 concentration and a reduction in the number of idling vehicles. The reduction in combustion emissions suggests that our dynamic messaging system can positively influence driver idling behavior in an area that typically experiences concentrated vehicle idling.

468b Diffusion Coefficients of Fullerene and Silica Nanoparticles (<10 nm) in Air By Fully Atomistic Molecular Dynamics.

Katerina S. Karadima, Dimitrios Tsalikis, Vlasios Mavrantzas, Sotiris E. Pratsinis

Given the advances in molecular dynamics and their impact on aerosol science (Mavrantzas and Pratsinis, 2019), a more accurate way to obtain diffusion coefficients for NPs with sizes less than 20 nm is through detailed atomistic simulations by employing accurate forcefields and shapes for the NPs and the surrounding gases. To this end, we have employed all-atom molecular dynamics (MD) simulations to study the diffusion coefficients of small fullerene and silica NPs with diameters from 0.4 to 7 nm in air at room temperature ($T = 298$ K) and pressure ($P = 1$ atm) by fully accounting for the diatomic shape and force field of air molecules (Tsalikis et al., 2023).

The MD results are averaged over several independent simulations to enhance sampling and decrease the statistical uncertainty of simulations. The MD-extracted diffusion coefficients are then compared to nearly all theoretical expressions widely used in the literature for NP diffusivity from the classic Cunningham-Millikan (CM) equation to the Fernández de la Mora et al. (2003) one, as well as that by the elementary kinetic theory of gases. We have also studied the diffusivity of a single silica molecule and several of the fullerene NPs modeled here as a function of temperature from 100 to 600 K.

The present MD-calculated diffusion coefficients show systematic deviations, already reported in the literature (e.g., by Fernández de la Mora et al., 2003, Li and Wang, 2003, Rudyak et al., 2002, Tammet, 1995), from the CM equation, especially as the NP diameter becomes comparable to the size of air molecules. This is justified because the CM equation was based mainly on experimental data with much larger particles at low pressures. Especially for the simulated NPs with diameter smaller than 2 nm, the diffusivity in air cannot be captured even by modifications of the CM equation that accounts for the size of gas molecules, as proposed by Tammet (1995) to account for the collision diameter and the transition from the elastic to inelastic collisions or by Fernández de la Mora et al. (2003) including an "effective" particle diameter instead of the true geometric particle diameter or even the reduced mass in a hybrid Epstein formula.

By comparing the present MD-derived diffusion coefficients with the above formulae and also the predictions of the kinetic theory of gases and the empirical formula by Fuller et al. (1966), we obtain a better understanding of how the NP size should be treated in these formulations. More specifically, we propose a new modification to the Cunningham-Millikan equation that captures quite well the data in the free molecular and transient regimes.

In addition, we provide new simulation data for the temperature dependence of the NP diffusivity in the free molecular and transient regimes, which allows us to evaluate existing scaling laws as well as to check if the exponents in these laws depend on NP size as has been reported by Rudyak et al. (2019).

References

J. Fernández de la Mora, L. de Juan, K. Liedtke, A. Schmidt-Ott, Mass and size determination of nanometer particles by means of mobility analysis and focused impaction, *J. Aerosol Sci.* **34**, (2003) 79-98, doi.org/10.1016/S0021-8502(02)00121-0.

N. Fuchs, The mechanics of aerosols, (Dover Publications Inc, New York, 1964).

S. Fuller, P.D. Schettler, J.C. Giddings, A new method for predictions of binary gas-phase diffusion coefficients, *Ind. Eng. Chem.* **58** (1966) 18-27, doi.org/10.1021/ie50677a007.

Z. Li, H. Wang, Drag force, diffusion coefficient, and electric mobility of small particles. I. Theory applicable to the free-molecule regime, *Phys. Rev. E* **68** (2003) 061206, doi.org/10.1103/PhysRevE.68.061206.

V.G. Mavrantzas, S.E. Pratsinis, The impact of molecular simulations in gas-phase manufacture of materials, *Curr. Opinion Chem. Eng.* **23** (2019) 174-183, doi.org/10.1016/j.coche.2019.04.006.

V.Ya. Rudyak, S.L. Krasnolutskii, A.G. Nasibulin, E.I. Kauppinen, Methods of measuring the diffusion coefficient and sizes of nanoparticles in a rarefied gas, *Dokl. Phys.* **47** (2002) 758-761, doi.org/10.1134/1.1519325.

V.Ya. Rudyak, S.N. Dubtsov, A.M. Baklanov, Measurements of the temperature dependent diffusion coefficient of nanoparticles in the range of 295–600 K at atmospheric pressure, *Aerosol Sci.* **40** (2019) 833-843, doi.org/10.1016/j.jaerosci.2009.06.006.

H. Tammet, Size and mobility of nanometer particles, clusters and ions, *J. Aerosol Sci.* **26** (1995) 459-475, doi.org/10.1016/0021-8502(94)00121-E.

D.G. Tsalikis, V.G. Mavrantzas, S.E. Pratsinis, Dynamics of molecular collisions in air and its mean free path, *Phys. Fluids.* **35**, (2023) 097131, doi.org/10.1063/5.0166283.

468c Uncertainty Quantification and Parameter Constraint in Complex Environmental Models: A Case Study of Smoke Influencing Radiative Balance in Africa

Victor Sanchez, Hamish Gordon, Mikael Kuusela, James Carzon

Development of environmental models is key to understanding future climate and informing related policy. Constructing methods for improving the accuracy of these models can be just as important as their initial development. Validating environmental models with observations and measurements is necessary to reduce uncertainties and identify shortcomings in a model's capabilities.

Here, we use a case study of seasonal wildfires in Southern Africa to develop methods for uncertainty quantification and constraint of input parameters in the UK Earth System Model, developed by the Met Office in the UK. This case study poses a unique problem because the Southern African wildfires (1) release the largest portion of the planet's biomass burning aerosols into the atmosphere and (2) the prevailing winds carry the smoke over the south Atlantic ocean where they interact with clouds. These wildfire emissions and their interactions with clouds have significant implications for the planet's radiation budget. Reducing large uncertainties associated with inputs to the model is key to developing a better understanding of the true magnitude of the importance of these seasonal fires in Southern Africa.

We constrain the uncertainty in the radiative effects of smoke from seasonal biomass burning in Southern Africa which we simulate in a climate model by considering the uncertainty in the input parameters to that model. We hypothesize that we can parameterize a large fraction of the uncertainty in smoke radiative effects by varying 12 uncertain parameters in the model. Some parameters scale emissions, both of smoke and of natural aerosols and precursors, and others scale parameters in uncertain processes such as the vertical velocities used to activate cloud droplets. Since satellites are not capable of measuring radiative effects, we start by comparing simulated aerosol optical depth (AOD) with satellite measured AOD to rule out combinations of parameters that produce implausible results. Prior work (e.g. Johnson et al, Atmos Chem Phys 2020) has shown AOD is capable of constraining model predictions involving the planet's radiative balance. We calculate the fire radiative effects by comparing simulations with and without smoke emissions. After AOD, we then apply the same methodology to simulated and satellite measured cloud droplet number concentration (CDNC) and cloud liquid water path (CLWP) to develop a stronger constraint on the radiative effect of these fires. Here, the variables for which values are compared between simulation and satellite will be called comparison variables.

To constrain our uncertain parameters, we built a perturbed parameter ensemble (PPE) consisting of 121 runs of the atmosphere-only UK Earth System Model spanning our space of 12 parameters and this was used as training data to build surrogate models. The values for the 121 parameter variants of the PPE were set using a Latin hypercube to ensure an adequate preliminary coverage of the parameter space. We simulate only the 2017 biomass burning season, and write data out from the simulations at a three-hourly time resolution, which allows us to understand the transport of individual smoke plumes. The surrogate models are trained using Gaussian Process Regression and allow us to approximately evaluate the model over the 12-D parameter space at a high coverage by utilizing 100,000 surrogate model parameter variants.

To rule out implausible parameter combinations that yield simulated comparison variables that are inconsistent with the observations, we use statistical methods developed in Carzon et al

(Environmental Data Science, 2023). These allow us to quantify uncertainty in the parameters at a high level of statistical confidence, improving on previous related work using the method of 'history matching' (whose theoretical guarantees are not well understood). We build on the statistical methods of Carzon et al. which enable principled constraints on uncertain parameters with guarantees of frequentist coverage. The methods developed here can be applied to uncertainty analysis and parameter constraint across multiple variables in other complex chemical and atmospheric models. We will present the methods, the constraints on parameters we have achieved to date, and the implications for the smoke radiative effects.

468d The Hygroscopicity of Nano-Plastic Particles and Implications for Cloud Formation and Climate

Chun-Ning Mao, Kanishk Gohil, Dewansh Rostogi, Akua Asa-Awuku

Nano-plastics have been found in snow and evidence shows that they can act as nuclei. The aerosol water-uptake ability is an important property that will determine wet deposition, a potential main route for nano-plastics to fall to the ground. In this work, we measure the hygroscopicity of three nano-plastics - low-density polyethylene (LDPE), polyethylene terephthalate (PET) and polyvinyl chloride (PVC), and a fourth compound, cellulose, with in- situ aerosol techniques via a cloud condensation nuclei counter (CCNC). Cellulose is a high- molecular weight compound and a main component of commonly used paper products. Ultrafine nano-plastic particles in the range of 20 to 200 nm are explored. Nano-plastic materials LDPE, PET, and PVC form droplets more readily than cellulose nanoparticles. The high hygroscopicity of nano-plastics indicates that the nano-plastics are either oligomers or polymer with properties for water adsorption. Furthermore, a single parameter hygroscopicity, κ, derived from different droplet activation models (Flory Huggins Köhler and Frenkel-Halsey-Hill adsorption theory) are reported. Moreover, the molecular weights of the leeched oligomers are predicted by Flory Huggins Köhler theory assuming ideal polymer mixing. Thus nano-plastics can be incorporated into clouds, transported globally, and may also be removed from the atmosphere via wet deposition processes. The observed intrinsic hygroscopicity is size-dependent and the adsorption-based hygroscopicity model could be applied to estimate the lifetime and transport of nano-plastic particles in the atmosphere.

468e Improvement of Surface PM2.5 Diurnal Variation Simulations in East Africa

Jun Wang, Chengzhe Li

The Multi-Angle Imager for Aerosols (MAIA), supported by NASA and the Italian Space Agency, is planned for launch into space in 2025. As part of its mission goal, outputs from a chemical transport model, the Unified Inputs for Weather Research and Forecasting Model coupled with Chemistry (UI-WRF-Chem), will be used together with satellite data and surface observations for estimating surface PM2.5. Here, we develop a method to improve UI-WRF-Chem with surface observations at the U.S. embassy in Ethiopia, one of MAIA's primary target areas in east Africa. The method inversely models the diurnal profile and amount of anthropogenic aerosol and trace gas emissions. Low-cost PurpleAir sensor data are used for validation after applying calibration functions obtained from the collocated data at the embassy. With the emission updates in UI-WRF-Chem, independent validation for February 2022 at several different PurpleAir sites shows an increase in the linear correlation coefficients from 0.1−0.7 to 0.6−0.9 between observations and simulations of the diurnal variation of surface PM2.5. Furthermore, even by using the emissions optimized for February 2021, the UI-WRF-Chem forecast for March 2022 is also improved. Annual update of monthly emissions via inverse modeling has the potential and is needed to improve MAIA's estimate of surface PM2.5.

468f Light Absorption Dynamics of Brown Carbon Particles during Wood Combustion and Pyrolysis

Constantinos Moularas, Philip Demokritou, Georgios Kelesidis

The light absorption dynamics of brown carbon (BrC) particles emitted during combustion or pyrolysis of pinewood are elucidated here in a tube furnace reactor connected with a variety of real-time monitoring and time-integrated sampling instrumentation. The BrC particles emitted by pinewood combustion contain about 80 % of condensed volatile organic compounds (VOCs), regardless of the O_2 concentration, $[O_2]$, in the tube furnace. Removing the condensed VOCs by thermal denuding reveals that BrC nanoparticles from wood pyrolysis have a bi-modal size distribution containing 95 % of small particles with a mean mobility diameter, d_m = 37 nm and 5 % of large particles with mean d_m = 107 nm. Increasing $[O_2]$ from 0 to 20 vol %, increases the fraction of large BrC nanoparticles up to 29 % and decreases their mean d_m to 78 nm. In this regard, the average mass absorption cross-section, *MAC*, of BrC increases from 0.1 to 0.27 m^2/g with increasing $[O_2]$. This indicates that the light absorption of BrC from wood combustion and pyrolysis is determined by the fraction of large particles with mean d_m = 78-107 nm. The BrC *MAC* measured here can be interfaced with global climate models (Kelesidis et al., 2022) to estimate accurately the contribution of particulate emissions from biomass combustors and wildfires to global warming.

Reference:

G.A. Kelesidis, D. Neubauer, L.-S. Fan, U. Lohmann, S.E. Pratsinis, Enhanced light absorption and radiative forcing by black carbon agglomerates, *Environ. Sci. Technol.* 56 (2022) 8610–8618.

528a Integration of Low-Grade Waste Heat in Direct Air Capture of CO_2 Systems

Lindsey Hamblin, Matthew D. Green, Klaus S. Lackner

Implementation of direct air capture technologies for the atmospheric removal of carbon dioxide are limited in economic feasibility due to the high costs of sorbent regeneration and upgrading the concentration of captured carbon dioxide. These costs can be mitigated through integration of direct air capture with industries that produce waste heat, utilizing this waste heat for the regeneration of sorbents. The research herein proposes a mechanism for upgrading the concentration of carbon dioxide and/or regenerating liquid sorbents with a series of stages, whereby each stage increases the partial pressure of carbon dioxide using vapor-liquid equilibria driven by the thermal gradient produced by integrated waste heat. Proof of concept of the design has been achieved using both modeling/simulation work and experimental data. Parameters including sorbent chemistry, temperature, flow rate, and feed CO_2 pressure are evaluated for impact on the overall output and efficiency of the system. Integration of the proposed DAC system in data centers is utilized as a case study.

528b Climate Action Tool (CAT): Philosophy and Case Studies

Ajay Koushik V, Ram Kishore Sankaralingam, Samir Paudel, Nakul Neupane, Sudharshan Saranathan, Dhruv Raghunath, Nishanth Nethaniel Magesh, Sumedh Kulkarni, Shvetha Sivaprasad, Gowri Shankar Navagana, Satyanarayanan Seshadri, Preeti Aghalayam, Raghunathan Rengaswamy

The pressing concern of climate change requires urgent attention, with rising carbon dioxide emissions from human activities being a major factor. The most common greenhouse gas, carbon dioxide, makes up around 65% of all greenhouse gas emissions. Carbon dioxide emissions have surged by around 90% since 1970, mostly due to industrial activities and the combustion of fossil fuels. So, it is crucial to monitor and mitigate the release of carbon dioxide into the atmosphere through various sectors.

To address this challenge effectively, a comprehensive and adaptable tool for assessing and strategizing carbon dioxide mitigation efforts is being developed. This tool aims to provide a robust approach that can be applied across different sectors and sub-sectors. By modelling sectors and sub-sectors using first principles, the tool offers a holistic view of emission sources. Any source related to CO_2 emission can be integrated as a model in the tool. At the core of this tool is a modular design that allows for customization and integration of sector-specific parameters and equations. This modular approach enables users to model emissions from any sector based on input and output data, with the flexibility to enhance the model with additional phenomenological or data-driven elements as needed.

Utilizing mass and energy conservation equations to control volumes within each sector and sub-sector, the tool measures carbon dioxide emissions concerning factors such as feed quality, energy sources, and process efficiency. This approach identifies the important parameters influencing emissions and offers a foundation to develop effective mitigation strategies at minimal investment. One key strength of this tool is its ability to offer policymakers and industry stakeholders actionable insights into emission reduction. The tool allows for the analysis of several low-carbon strategies by incorporating optimization approaches to reduce the cost and emissions of cross-sectoral systems. Broadly, it facilitates the generation of strategies that require minimal targeted investments, thereby leading to considerably major emission reductions. For instance, to meet certain emission reduction targets given by government in a country, this tool can suggest the best technically and economically viable decarbonization strategy, such as an optimal fuel-mix ratio, carbon capture and utilization (CCU) along with the best timing for installation, or a combination thereof.

In one of our studies using the tool, we interconnected different industrial sectors: cement, steel, power, and hydrogen, where the latter two industries meet the demand of the former two industries. Intending to minimize the total cost (CapEx and OpEx) and total CO_2 emissions, with limits on fuel availability and technological constraints, we created two scenarios to analyze with this tool: (i) Current Landscape, (ii) Transformative Path Forward. In the first scenario without additional capital investment, only fuel-mix was optimized within the existing infrastructure. This showed heavy reliance on fossil fuels across sectors, with substantial investments needed in each sector. For the second scenario, the tool suggested to focus on the electrification of steel industry, good energy-mix and installation of CCS plant in cement and power industries in 2nd year and also in fossil-based hydrogen production unit in 4th year. While comparing between cement and steel sector, the tool showed that cement sector needs more investment due to fossil-based energy mix and mandatory CCS for process emissions. Scenario-2 used 25% less coal and 65% less natural gas as compared to

Scenario-1. A strategic energy mix, featuring electrification and CCS, played a key role in achieving projected impressive 83% reduction in CO_2 emissions. It is important to emphasize that achieving such significant reductions in emissions across various sectors required minimal investment, which would have been substantially higher if these sectors were analysed individually. These types of analyses can aid policymakers, and industry stakeholders by providing an estimate of the budgets required to build strategies for decarbonization. Similarly, many such scenarios pertaining to agriculture, manufacturing, and transport can be generated, studied and interesting insights can be obtained using this tool.

Recent literature has introduced various tools designed to support policymakers and stakeholders in reducing emissions. Wang et al. [1] developed a model that explores different decarbonization strategies within China's power sector, considering spatial and temporal dynamics. Li et al. [2] investigated how electrification could lower emissions from China's light-duty trucks. Tan et al. [3] introduced a model that facilitates collaboration between government and industry to address maritime emissions, with the government suggesting CO2 reduction methods and industries implementing feasible solutions with minimal financial burden and reduced penalties. This model emphasizes a subsidy-penalty mechanism to achieve optimal decarbonization outcomes. Matamala et al. [4] examined the role of carbon capture and storage (CCS) technologies, coupled with negative emission techniques like biomass, in Latin America's power sector. Hebada et al. [5] explored various paths towards achieving net-zero emissions in Brazil's iron and steel industries by 2050. Vernay et al. [6] created a model to assess the effects of energy transitions on energy community businesses. In contrast to these existing models and tools, this tool presents numerous unique advantages. While reported models in the literature may be confined to specific sectors or regions, the proposed tool with a user-friendly interface and customizable functionalities, is applicable across various sectors and regions. Moreover, its basis in first principles ensures robustness and reliability. This feature not only enhances usability but also facilitates well-informed decision-making for policymakers and industry stakeholders.

Thus, the development of this comprehensive tool offers a versatile and reliable solution for addressing carbon dioxide emissions across different sectors and sub-sectors worldwide. Its adaptable design, rooted in fundamental principles, representing a significant leap forward in environmental management, delivers practical insights for policymakers and industry stakeholders to devise effective mitigation strategies, towards achieving emission reduction targets and fostering a sustainable, low-carbon economy, thereby contributing to the global endeavor to combat climate change.

References

[1] Y. Wang, Z. Zhao, W. Wang, D. Streimikiene, T. Balezentis, Interplay of multiple factors behind decarbonization of thermal electricity generation: A novel decomposition model, Technological Forecasting and Social Change. 189 (2023). https://doi.org/10.1016/j.techfore.2023.122368.

[2] J. Li, L. Wang, C. Ji, H. Liu, L. Lu, T. Zhang, M. Zhao, S. Xu, Assessing the decarbonization potential of China's light-duty truck fleet by electrification, Energy Reports. 9 (2023) 212–225. https://doi.org/10.1016/j.egyr.2023.03.018.

[3] R.R. Tan, I.H. V. Gue, J.F.D. Tapia, K.B. Aviso, Bilevel optimization model for maritime emissions reduction, Journal of Cleaner Production. 398 (2023). https://doi.org/10.1016/j.jclepro.2023.136589.

[4] Y. Matamala, F. Flores, A. Arriet, Z. Khan, F. Feijoo, Probabilistic feasibility assessment of sequestration reliance for climate targets, energy. 272 (2023) 127160.

https://doi.org/10.1016/j.energy.2023.127160.

[5] O. Hebeda, B.S. Guimarães, G. Cretton-Souza, E.L. La Rovere, A.O. Pereira, Pathways for deep decarbonization of the Brazilian iron and steel industry, Journal of Cleaner Production. 401 (2023) 136675. https://doi.org/10.1016/j.jclepro.2023.136675.

[6] A.L. Vernay, C. Sebi, F. Arroyo, Energy community business models and their impact on the energy transition: Lessons learnt from France, Energy Policy. 175 (2023). https://doi.org/10.1016/j.enpol.2023.113473.

528d Advancements in Direct Air Capture: Unveiling a Simple and Robust Synthesized Fibrous Amine-Functionalized Matrix (FAM) Sorbent for Commercial Scale-up

Qiuming Wang, Walter Wilfong, Fan Shi, McMahan Gray

The increasing urgency to address global climate change has driven extensive research into carbon capture and sequestration (CCS) technologies. Among these, direct air capture (DAC) has emerged as a pivotal solution due to its potential to directly remove CO_2 from the atmosphere. DAC operates independently of emission sources, offering flexibility in location and the potential to significantly reduce atmosphere CO_2 concentrations. However, the effectiveness and scalability of DAC technologies hinge on the development of efficient and economically viable sorbent materials. The quest for an effective sorbent focuses on materials that can selectively and efficiently capture CO_2 from ambient air, are durable over many capture and release cycles, and can be produced and operated at low costs. Recent advancements in material science have led to the exploration of amine-functionalized sorbents, hollow fiber sorbents and membranes, and other novel materials designed to meet these criteria. The development of a simple and robust synthesized Fibrous Amine-functionalized Matrix (FAM) sorbent represents a significant stride in this direction. The FAM can be efficiently functionalized onto substate via a dip-coating process involving immersion in the formula solution for several seconds, followed by drying at 80-90 °C for 1 hour. This method facilitates straightforward and cost-effective scalability. The viability of this approach was demonstrated through the preparation of a 1-meter FAM roll on a laboratory scale, achieved manually. Furthermore, the FAM exhibits a dry DAC capacity of ~ 1.7 mmol/g-FAM under 5 hours equilibrium, a performance that stands out within the spectrum of reported sorbent materials. In addition, the FAM demonstrated a release of over 90% CO_2 at 90 °C within 20 minutes, indicating low energy requirements for regeneration and high efficiency. The FAM sorbent offers the potential for high CO_2 capture efficiency, low energy requirements for CO_2 release, and suitability for commercial scale-up. All of these help to address key challenges in the deployment of DAC technology.

528f Techno-Economic and Life-Cycle Analysis of Integrated Carbon Capture and Conversion to Materials (IC3M) Platform for Producing Synthetic Natural Gas and Methanol

Yuan Jiang, Jothi Kothandaraman, Robert A. Dagle, David J. Heldebrant

Carbon capture and conversion is a strategic approach to alleviate climate change by simultaneously reducing greenhouse gas (GHG) emissions and generating value-added products from captured carbon. Over the past decade, several demonstration plants have been constructed in Europe to produce synthetic natural gas (SNG) or methanol from CO_2 (Jiang et al., 2024). Each plant includes a carbon capture unit to generate pure CO_2 from industrial exhaust gas at low pressure, a CO_2 compressor to bring CO_2 to a high pressure required by downstream reactions and a thermochemical CO_2 hydrogenation unit to convert CO_2. However, these plants, which use a Separated Carbon Capture and Conversion to Material (SC3M) platform, are often energy-intensive, mainly due to the solvent/sorbent regeneration and CO_2 compression steps, resulting in a high conversion cost. Leveraging emerging water-lean solvents developed for low-cost carbon capture (Jiang et al., 2023), recently, an integrated carbon capture and conversion to material (IC3M) platform has been developed at Pacific Northwest National Laboratory (Kothandaraman et al., 2022), and proven in a lab-scale continuous flow system. In the IC3M platform, CO_2 in the flue gas is captured by water-lean solvent. This CO_2-rich solvent then directly reacts with H_2 in a condensed phase, while the energy-intensive solvent regeneration and CO_2 compression steps are eliminated. Consequently, it is anticipated that the energy consumption and cost of the IC3M platform will be significantly lower than those of the SC3M platform, paving the way for a more sustainable future. In this talk, we will present process models developed in Aspen Plus for IC3M technologies to produce both SNG and methanol based on experimental data, and a comprehensive techno-economic and life-cycle analyses to compare IC3M, SC3M, and fossil-based methanol and methane. The key measures include, but are not limited to, energy efficiency, minimum product selling price, conversion cost, and cradle-to-grave GHG emissions.

References

Jiang et al., 2024. Emerging technology review: carbon capture and conversion to methane and methanol. Technical Report. PNNL-35588. Pacific Northwest National Laboratory, Richland, WA.

Jiang et al., 2023 Energy-effective and low-cost carbon capture from point-sources enabled by water-lean solvents. Journal of Cleaner Production 388, 135696.

320a Intensified Direct Air Capture Process Using Aqueous Amino Acid Solvents with Low Regeneration Temperature

Gyoung Gug Jang, Abishek Kasturi, Aye Meyer, Dhruba Jyoti Deka, Diana Stamberga, Radu Custelcean, Costas Tsouris

Aqueous amino acids, such as potassium sarcosinate (K-SAR), have emerged as high-performance direct air capture (DAC) solvents due to their low volatility, high CO_2 capacity, rapid CO_2 uptake kinetics, low-temperature and energy of regeneration requirements at 110 °C, and because they are environmentally friendly solvents. Our study focuses on enhancing DAC efficiency through an intensified process employing a high-flux air-liquid contactor, catalytic regeneration and sub-ambient solvent formulation to improve DAC plant economics. We developed a low-cost, corrosion-resistant high-flux DAC (HiDAC) contactor exhibiting excellent compatibility with K-SAR, showcasing remarkable CO_2 capture rates and efficiencies within a compact configuration. Subsequently, catalytic regeneration was adopted to reduce the energy burden of solvent regeneration. A TiO_2 solid catalyst demonstrated significant improvements in CO_2 desorption rates and cumulative CO_2 removal at lower regeneration temperatures, leading to a substantial reduction in the overall regeneration energy requirements due to reductions in the sensible and vaporization heats. To address potential sub-freezing DAC conditions, we applied anti-freezing solvent formulations and investigated the performance of K-SAR at sub-freezing temperatures. We explored the efficacy of ethylene glycol additives in preventing solvent freezing, revealing promising CO_2 fluxes across a wide temperature range. Finally, our study offers a comprehensive technoeconomic analysis of these intensified DAC approaches, providing insights into their feasibility and potential for widespread adoption. By integrating advanced materials and innovative processes, we aim to contribute to the development of more sustainable and cost-effective DAC solutions, crucial for combating climate change.

320b Sub-Ambient CO_2 Flux Measurements and Modeling Using Amino Acid Solvents

Abishek Kasturi, Gyoung Gug Jang, Jorge Gabitto, Diana Stamberga, Jiho Seo, David Sholl, Radu Custelcean, Sotira Yiacoumi, Costas Tsouris

One of the main advantages of direct air capture (DAC) over point-source capture is the flexibility in plant siting anywhere around the world. In this study, aqueous amino acid solutions, specifically those of potassium sarcosinate (K-SAR), were used to capture CO_2 from air. Because plant siting involves areas where the temperature is occasionally below the freezing point of water, termed here as sub-ambient temperatures, there was a need to investigate the physical properties and performance of the solvent at such temperatures. The need for anti-freezing additives, such as triethylene glycol, was determined by comparing the freezing point, CO_2 flux, and mass-transfer coefficient of K-SAR alone and K-SAR with glycol. The solvent viscosity and density were also measured for different temperatures and solvent concentrations. A reactor was set up to reliably measure the CO_2 flux at different temperatures by ensuring a constant, measurable surface area and effective temperature control. The CO_2 flux was measured between -20 and 45 °C for 1 and 3 M K-SAR, and the results were compared to theoretical estimations based on the solvent's physical properties and operating conditions. For 1 M K-SAR at 25°C, the flux was experimentally determined at 5.9×10^{-5} mol $m^{-2}s^{-1}$, while the theoretical flux was estimated at 5.4×10^{-5} mol $m^{-2}s^{-1}$, showing good agreement between experimental data and theoretical results. The results of this study provide the fundamental basis for developing an effective carbon capture technology with amino acid solvents that can be used to effectively capture atmospheric CO_2 at scale in different climates.

320c Enhancing CO_2 Capture Efficiency: Cordierite Monoliths Coated with Zeolite 13X Under Microwave-Assisted Heating Conditions

Matheus Strobel, Mustafa Erguvan, Shahriar Amini

The urgency to address global climate change, fueled by greenhouse gas emissions, is intensifying. Rapidly increasing in atmospheric CO_2 levels is the main concern. Current predictions show a substantial increase in CO_2 emissions, which could lead to a rise in global temperatures by 1.4 to 6.1 C by the end of the century [1]. The development of efficient gas–solid contactors is crucial for advancing CO_2 capture technologies, particularly in applications such as post-combustion carbon capture and direct air capture (DAC). Monolithic structures present an appealing alternative to conventionally prepared catalyst pellets or powders, boasting several superior properties. They excel in mass transfer efficiency, maintain low pressure drop, exhibit thermal stability, and demonstrate robust mechanical strength compared to traditional catalyst pellets or powders [2]. Monoliths are basically uniform blocks, consisting of parallel channels that can be extruded into different shapes and sizes. Wash coating by slurry coating method and in situ synthesis are the most industrially known processes applied to the monolith to improve surface properties to increase desorption capacity.

Microwave-assisted heating (MW) is one of the emerging technologies for post combustion carbon capture and DAC applications. It is claimed to be energy efficient, time-saving, and improves throughput compared to conventional processing systems, due to concentration heat capacity [3].

In this study, a cordierite monolith wash coated with Zeolite 13X has been used for CO_2 capture of post combustion under mono-mode microwave conditions. The wash coating process involved formulating a suspension of Zeolite 13X and deionized water, followed by immersion of mini monoliths, air drying, and calcination. For zeolite synthesis, a synthesis solution was prepared and introduced with cordierite mini monoliths into an autoclave for crystallization. The in-situ synthesis method yielded monoliths with superior characteristics compared to slurry-coated counterparts, showcasing enhanced mechanical stability and homogeneous zeolite layers, thus offering promising prospects for practical CO_2capture applications.

Microwaves provide precise control, allowing for accurate temperature regulation and uniform heating throughout the sorbent material. The monolithic sorbents were thoroughly characterized and compared based on different coating layers to investigate its effect on the desorption characteristics. Adsorption and microwave desorption experiments were also tested varying two key parameters: regeneration temperature and microwave power. Based on the comparison of the two different methods investigated under experimental conditions, like microwave power and reactor temperature, impact CO_2 adsorption on slurry-coated monoliths.

References

[1] X. Shi, H. Xiao, H. Azarabadi, J. Song, X. Wu, X. Chen, K.S. Lackner, Sorbents for the Direct Capture of CO2 from Ambient Air, Angewandte Chemie International Edition, 59(18) (2020) 6984-7006.

[2] T.A. Nijhuis, A.E.W. Beers, T. Vergunst, I. Hoek, F. Kapteijn, J.A. Moulijn, Preparation of

monolithic catalysts, Catalysis Reviews, 43(4) (2001) 345-380.

[3] I.J. Siddique, A.A. Salema, E. Antunes, R. Vinu, Technical challenges in scaling up the microwave technology for biomass processing, Renewable and Sustainable Energy Reviews, 153 (2022) 111767.

320d Guanidine-Based Adsorbents for CO_2 Capture

Xakin Ramirez Isunza, Brandon Tapia, Stephen Martin

Over the past 200 years—correlated with the beginning of the industrial revolution—atmospheric carbon dioxide (CO_2) concentration has increased as a direct result of human activity. Current CO2 concentrations are around 420 parts-per-million (ppm). The accumulation of CO_2 and other greenhouse gases in the atmosphere has increased global average temperatures, leading to increased severity and frequency of extreme weather and climate events. To meet the Paris Agreement objective of keeping the average global temperature under 2 °C above pre-industrial temperature, removal of 10 billion metric tons per year of atmospheric CO_2 must occur before 2050 with another 20 billion metric tons per year after 2050. One of several negative emissions technologies proposed to achieve this aim is direct air capture (DAC) of CO_2. One approach to DAC that aims to lower the energy demand of the DAC process is the use of solid adsorbents.

Of the numerous solid sorbent classes proposed for CO_2 capture, such as amine-modified porous materials, zeolites, metal-organic frameworks, covalent-organic frameworks, and polymers, amine-modified materials are interesting because of their high selectivity and capacity even at DAC conditions as they have the ability to perform physisorption within the pores and chemisorption due to the acid-base interactions between amine groups and CO_2.

Here we present a series of activated carbon adsorbents functionalized with guanidine carbonate *via* wet impregnation with the objective of improving the CO2 adsorption capacity of activated carbon in low CO2 concentrations and in the presence of humidity. We performed CO_2 breakthrough tests under DAC conditions (420 ppm CO_2, 22 °C, 1 bar) on these adsorbents and found that the addition of guanidine carbonate at all tested loading levels improves the CO_2 adsorption capacity of activated carbon, and the presence of humidity improves CO_2 uptake further. The observed optimal guanidine carbonate loading level was 10 wt %.

320e Influence of Metal Contaminants on Solid Amine Sorbents for Direct Air Capture

Botagoz Kuspangaliyeva, Hyun June Moon, Yoseph Guta, Akriti Sarswat, Christopher W. Jones, Ryan Lively

Direct Air Capture (DAC) of CO_2 from the air via adsorption is a promising technology to address current and historic CO_2 emissions. Amine-containing sorbents have demonstrated potential for capturing CO_2 from ambient air due to their high CO_2 adsorption selectivity. However, existing research on amine-based DAC adsorbents has been primarily conducted under ideal conditions with clean feed streams, neglecting the influence of impurities present in real-world applications. Contaminants may strongly interact with the surface functional groups of adsorbents, affecting their performance and structural integrity. Metal ion species are of particular importance as potential impurities, as even trace amounts may catalyze amine oxidation. This study aims to explore the influence of transition metal species such as Cu^{2+}, Fe^{2+}, and Ni^{2+} on the CO_2 uptake performance and oxidative degradation of an aminopolymer-impregnated silica sorbent (30 wt.% poly(ethyleneimine) (PEI)/SBA-15). Our ongoing research investigates the effect of various metal-loading concentrations on the CO_2 adsorption capacity of both fresh and oxidized adsorbents, employing IR spectroscopy to gain deeper insights into the catalytic effect of the metals on amine oxidation. We observe that metals can deactivate the adsorbent by poisoning the available amine sites. Moreover, exposure of the adsorbent to air at elevated temperatures demonstrates that the presence of metals significantly accelerates amine oxidative degradation that seems to follow a radical-initiated autoxidation mechanism.

320f Quantification of Degradation of Amine Solvent in CO_2 Capture Process.

Abhimanyu Singh Khichi, Deeksha Sundarrajan, Swapna Singha Rabha

Extreme weather caused by global warming nowadays is occurring more frequently on a worldwide scale. Global warming resulting from the emission of greenhouse gases has received widespread attention. Among the greenhouse gases, CO_2 is the major contributor to global warming. The primary stationary CO_2 sources are coal-fired and natural gas-fired power plants. Therefore, CCS technologies must be developed to cope with global warming and climate change due to increasing CO_2 emissions. Absorption of CO_2 using amine-based solvents has been considered highly promising because of their high rate and selectivity towards CO_2. However, primary and secondary amine solvents have significant drawbacks, such as high energy consumption during solvent regeneration, high oxidative and thermal degradation of solvent, high equipment corrosion, and low absorption capacity. Solvent degradation is an undesirable operating problem caused by prolonged exposure of amine solvents to CO_2, SOx, and oxygen in the flue gases, resulting in solvent loss and equipment corrosion. Therefore, it very important to quantify the factors influencing the formation of the degradation compounds.

In this work, the CO_2 absorption/desorption performance of aqueous Monoethanolamine (MEA) was studied. The experiments has been conducted in a screening apparatus, which consist of glass round bottom flask, condenser and a temperature control water bath. The 30wt% MEA solution was prepared using deionized water and verified by titration with 1 N standard HCl solution with methyl orange as an indicator. The experiments were conducted with both synthetic flue gases (N_2: CO_2) and (N_2: CO_2: O_2) at a flow rate of 300ml/min. The concentration of CO_2 in the mixed gas stream at the outlet was determined using a CO_2 gas analyzer. The absorption starts when the first bubble of the feed gas emerges into the amine solution. To determine CO_2 loading in amine solution and amine concentration, the liquid sample was taken at regular intervals until the process reached equilibrium. The CO_2 loading and amine concentration were determined using the Chittick apparatus by titration with 1N HCl by neutralization reaction and methyl orange as an indicator. The pH of the solution at regular intervals was also determined using the pH meter. The CO_2-rich amine solution from the absorption experiment was transferred to the desorption flask for the desorption process at 393 K using nitrogen gas for stripping out CO_2. In this work, the applicability of the experiment setup was verified by measuring the equilibrium solubility of CO_2 in 30wt% MEA and pH values, and compared with the literature data. The liquid sample collected at different time interval was further sent for analytical analysis using GC-MS and ICP.

The CO_2 absorption rate, pH values of 30wt% MEA solution at 313 K were measured at different time with a gas mixture flow rate of 300 ml/min and an inlet CO_2 mole fraction of 0.2. The pH of the initial 30wt% amine solution was 11.8 and it decreased to the value of 8 after injecting CO_2. The 30wt% MEA solution achieved the equilibrium CO_2 loading value of 0.55 mol CO_2 per mol amine after six hours of absorption experiment at 313 K, with a gas mixture flow rate of 300 ml/min and an inlet CO_2 mole fraction of 0.2. The above results agree with the literature data (Vega et al., 2018). The presence of degraded products in the solution was confirmed by GC-MS and HPLC. Further, the effect of concentration of O_2 and desorption temperature will be studied and its effect on the CO_2 absorption rate, CO_2 loading and amine degradation rate will be studied.

References

1. Vega, M. Cano, S. Camino, B. Navarrete, J.A. Camino, Evaluation of the absorption performance of amine-based solvents for CO2 capture based on partial oxy-combustion approach, International Journal of Greenhouse Gas Control, Volume 73, 2018, Pages 95-103, ISSN 1750-5836, https://doi.org/10.1016/j.ijggc.2018.04.005.

401a Towards Modelling and Structural Optimization of Adsorption-Based Direct Air Capture Technologies*

Hector Alejandro Pedrozo, Mayra Gonzalez Ramirez, Tiras Lin, Thomas Roy, Thomas Moore, Du Nguyen, Pratanu Roy, Sarah Baker, Lorenz Biegler, Grigorios Panagakos

Efforts within the scientific community to counteract CO_2 emissions and curb global temperature rise below 1.5°C above pre-industrial levels are driving the development of Carbon Dioxide Removal (CDR) strategies. Within CDR, Carbon Capture and Sequestration (CCS) presents a promising framework, however, Direct Air Capture (DAC) is crucial for addressing hard-to-avoid emissions. Scaling up these approaches requires collaborative endeavors among industry, stakeholders, and policymakers. Challenges such as high energy consumption and scalability hinder the full potential of DAC. Hence, innovation in algorithm development and computational tools focused on optimizing adsorption-based systems becomes crucial. Computational Fluid Dynamics (CFD) is a powerful tool that can quantitatively predict the effects of design parametrization on device performance. Within this framework, we advocate for the utilization of topology optimization, a versatile, flexible, and potent mathematical tool. Together, Topology Optimization and CFD create a promising framework to enhance devices and get closer to commercial scale [1]. To make this connection possible, it is imperative to first obtain detailed CFD models that can describe a DAC process to make decisions.

Over the last few years, DAC research has been mainly focused on sorbent development. To contribute to this, there is a need for detailed simulation that opens the room for optimization techniques. A detailed DAC system considers complex kinetics, heat transfer, and temperature dependence. Another challenge in deriving optimal adsorption systems is the inherent dynamic nature of the process and the not well-understood mass transfer between solid material and fluid. Furthermore, it is imperative to account for interactions between the packing material (such as pellets) and bulk flow within the system. To model this, we rely on previous studies that model mass transfer as a linear driving force alongside the relevant governing and constitutive equations and make use of stepped isotherms to describe the proper behavior [2].

To achieve our objectives, we initially set up a 3D system to describe a DAC adsorption step and incorporate kinetics, obtained from the relevant literature on cutting-edge materials for DAC (e.g., [2, 3, 4]). This dynamic model can be conceptualized as a pseudo-4D simulation where we describe the adsorption step in a 3D system and then add a 1D concentration profile along the packing pellet's radius. This coupled model is carried out on COMSOL Multiphysics®. The model requires a coupled, three-dimensional implementation of heat, momentum and mass transfer, and is therefore computationally expensive. Furthermore, it includes stiff ODEs for the adsorption kinetics, imposing model convergence limitations. To address this issue, we get inspiration from previous work in analytical solutions for isotherm models [5]. According to this approach, a reformulation is proposed that substitutes stiff ODEs with 1D analytical expressions for adsorption. Once the model is validated, it is translated to a 2D approach that will be useful within a structural optimization framework. The designs derived from the topology optimization/CFD approach serve as a starting point towards the development of large-scale, intricately detailed systems optimized for scalability.

References

[1] F. Okkels and H. Bruus, 'Scaling behavior of optimally structured catalytic microfluidic reactors,' Physiscal Review E, 2007.

[2] L. A. Darunte, T. Sen, C. Bhawanani, K. S. Walton, D. S. Sholl, M. J. Realff and C. W. Jones, 'Moving Beyond Adsorption Capacity in Design of Adsorbents for CO2 Capture from Ultradilute Feeds: Kinetics of CO2 Adsorption in Materials with Stepped Isotherms,' Industrial & Engineering Chemistry Research, pp. 366-377, 2019.

[3] R. Hughes, G. Kotamreddy, A. Ostace, D. Bhattacharyya, R. L. Siegelman, S. T. Parker, S. A. Didas, J. R. Long, B. Omell and M. Matuszewski, 'Isotherm, Kinetic, Process Modeling, and Techno-Economic Analysis of a Diamine-Appended Metal–Organic Framework for CO2 Capture Using Fixed Bed Contactors,' Energy Fuels, vol. 35 , no. (7), p. 6040–6055. , 2021.

[4] S.-m. Kinm and G. Léonard, 'Process modelling of Direct Air Capture (DAC) of CO2 using solid amine sorbents,' Computer Aided Chemical Engineering, pp. 265-270, 2022.

[5] P. Roy, S. T. Castonguay, J. M. Knipe, Y. Sun, E. A. Glascoe and H. N. Sharma, 'Multi-material modeling of sorption-desorption processes with experimental validation,' Chemical Engineering Science, vol. 253, 2022.

*This work was performed under the auspices of the U.S. Department of Energy by Lawrence Livermore National Laboratory under Contract DE-AC52-07NA27344 and was supported by Laboratory Directed Research and Development funding under project 22-SI-006. LLNL Release Number: LLNL-PROC-847581.

401b Design and Operation of Direct Air Capture (DAC) Systems Under Varying Environmental Conditions

Andrew Bai, Youn Ji Min, Jinsu Kim, Hannah Holmes, Matthew Realff

To address climate change caused by rising levels of carbon dioxide (CO_2) in the atmosphere, direct air capture (DAC) as a negative emission technology, has become an active area of research [1][2]. Several DAC sorbent-loaded fibers systems have been developed using different adsorbent materials. Adsorption on solid amine sorbent-loaded fibers is a promising alternative because of low manufacturing cost, low pressure drop, and use of low temperature steam for regeneration [3][4].

The objective of this work is to model, simulate, and optimize DAC systems utilizing silica-supported polyethylenimine adsorbents embedded in porous polymer fibers under varying ambient conditions. Literature studies have frequently assumed that the DAC process is run under constant lab conditions. However, the ambient temperature and humidity of many locations vary both seasonally and with the time of day, which can lead to different optimal operating conditions at different timescales [5]. In our approach the design is performed by optimizing the operation of a cycle over a limited number of scenarios where the design variable is the system size to capture a fixed quantity of CO_2. The scenarios are generated from statistical distributions of ambient temperature and humidity for a given location. To adjust the performance of the system to reflect the predicted temperature and humidity trends over several hours, dynamic optimization of operation is performed for several cycles using a rolling-horizon approach with surrogate models built from more detailed dynamic models used to perform optimization of the future cycles.

References:

1. McQueen, K. Vaz Gomes, C. McCormick, K. Blumanthal, M. Pisciotta, and J. Wilcox, 'A review of direct air capture (DAC): scaling up commercial technologies and innovating for the future,' in Prog. Energy, vol. 3, no. 3, Art. no. 032001, Apr. 2021, doi: 10.1088/2516-1083/abf1ce.
2. Pacala, M. Al-Kaisi, M. Barteau, et al., 'Negative Emissions Technologies and Reliable Sequestration: A Research Agenda,' in *Proc. Natl. Acad. Sci. U.S.A.*, 2019, doi: 10.17226/25259.
3. Rim, F. Kong, M. Song, C. Rosu, P. Priyadarshini, R. P. Lively and C. W. Jones, 'Sub-ambient temperature direct air capture of CO2 using amine-impregnated MIL-101(CR) enables ambient temperature co2 recovery,' JACS Au, vol. 2, no. 2, pp. 380-393, 2022.
4. Kalyanaraman, Y. Fan, R. P. Lively, W. J. Koros, C. W. Jones, M. J. Realff and Y. Kawajiri, 'Modeling and experimental validation of carbon dioxide sorption on hollow fibers loaded with silica-supported poly(ethylenimine),' Chemical Engineering Journal, vol. 259, pp. 737-751, 2015.
5. F. Wiegner, A. Grimm, L. Weimann, and M. Gazzani, 'Optimal Design and Operation of Solid Sorbent Direct Air Capture Processes at Varying Ambient Conditions,' Ind. Eng. Chem. Res., vol. 61, no. 34, pp. 12649–12667, Aug. 2022, doi: 10.1021/acs.iecr.2c00681.

401c CO_2 Adsorption Equilibrium of Modified Amine-Based Sorbent for Direct Air Capture

Joo-Youp Lee, James Akinjide

Direct air capture (DAC) continues to draw a lot of attention as it has been identified as a major technology for net-zero emissions control. Since it involves CO_2 separation directly from the atmosphere, there is a wide range of location options for deploying DAC plants under different climatic conditions. Therefore, it is important to understand the adsorption characteristics of a sorbent under different temperature and humidity conditions. However, to the best of our knowledge, there is a limited number of previous studies on adsorption isotherms under very dilute CO_2 concentrations relevant to DAC.

In this study, equilibrium adsorption data were obtained under ambient and sub-ambient CO_2 concentrations for unmodified and modified amine-based sorbents and described using the Toth isotherm model. Modified amine-based sorbents can reduce the shortcomings of thermal and oxidative degradations of unmodified amine-based sorbents. The isotherm data for modified amine sorbents showed a shift in equilibrium resulting from the amine modification, leading to a decrease in the heat of CO_2 adsorption. The modification most likely benefits from lowering the basicity and increasing the steric hindrance of the amine center, leading to lower binding energy between adsorbed CO_2 molecules and modified amine sites. Our differential scanning calorimetry (DSC) data for saturated sorbents also support that modified amine-based sorbents show lower heat of desorption than that of unmodified amine sorbent. These data support that modified amine-based sorbents can provide lower heat of desorption as well as the resistances to thermal and oxidative degradations.

Our DAC system includes a sorbent-washcoated monolith structure. The isotherm model determined from lab-scale testing of the powdered sorbent is a key part which will be employed in the coating layer of the monolith to predict the adsorption performance. This is done under various flow conditions within the system. Adsorption isotherm parameters determined from this effort provide a means to successful scale-up of DAC following key performance metrics like CO_2 capture efficiency, overall volumetric productivity, and pressure drop.

401e Operation of Direct Air Capture Unit Under Degradation Effects

Andrew Bai, Hannah Holmes, Matthew Realff

Recently, Direct air capture (DAC) has been a significant pathway to achieve negative emissions, and solid sorbent processes stand out thanks to relatively lower energy required for regeneration. Most active research has assumed constant behavior throughout the sorbent operational life. However, many sorbents under consideration for DAC degrade over operation time, which affects the equilibrium and kinetics under both the adsorption and desorption condition and causes reduction in the swing capacity over cycles if the operating conditions are constant [1][2]. Therefore, taking the degradation of materials into account is essential to simulate real DAC unit operation and to design effective long term operating strategies and material replacement timing.

Materials might have multiple degradation mechanisms, such as physical, thermal, and oxidative degradation, and dissolution leading to physical loss of adsorbent capacity [3] [4]. The degradation mechanisms are classified as reversible and irreversible, and by how the state trajectory of the adsorption or desorption impacts the rate of degradation. In addition, there may be changes in the support material properties such as loss of porosity which could impact the overall sorbent performance even without the sorbent itself degrading [5]. All classifications of degradation will cause cyclic productivity reduction, either permanently, reversibly, or partially reversibly, but have different forms of degradation over time.

In this work, mathematical models of material degradation for different degradation mechanisms have been constructed, and these models are embedded in the simulation model as a lumped degradation multiplier of the kinetic model or the equilibrium model respect to the effects. Then, the cycle parameters are adjusted based on predictions and state observations to account for the degradation. The final goal is to schedule optimally material regeneration for reversible degradation mechanisms, and schedule material replacement for irreversible and partially reversible degradation mechanisms.

Reference:

1. Shi, H. Xiao, H. Azarabadi, J. Song, X. Wu, X. Chen, and K. S. Lackner, 'Sorbents for the Direct Capture of CO2 from Ambient Air,' *Angew. Chem. Int. Ed.*, first published on 05 August 2019, doi: 10.1002/anie.201906756.
2. E. Holmes, S. Banerjee, A. Vallace, R. P. Lively, C. W. Jones, and M. J. Realff, 'Tuning sorbent properties to reduce the cost of direct air capture,' March 08, 2024, doi: 10.26434/chemrxiv-2024-8sc71.
3. Rosu, S. H. Pang, A. R. Sujan, M. A. Sakwa-Novak, E. W. Ping, and C. W. Jones, 'Effect of Extended Aging and Oxidation on Linear Poly(propylenimine)-Mesoporous Silica Composites for CO2 Capture from Simulated Air and Flue Gas Streams,' *ACS Applied Materials & Interfaces*, vol. 12, no. 34, 2020, doi: 10.1021/acsami.0c11185.
4. Miao, Y. Wang, X. Zhu, W. Chen, Z. He, L. Yu, and J. Li, 'Minimizing the effect of oxygen on supported polyamine for direct air capture,' *Separ. Purif. Technol.*, vol. 298, 121583, Oct. 2022, doi: 10.1016/j.seppur.2022.121583.
5. A. Saenz Cavazos, E. Hunter-Sellars, P. Iacomi, S. R. McIntyre, D. Danaci, and D. R. Williams, 'Evaluating solid sorbents for CO2 capture: linking material properties and process efficiency via adsorption performance,' *Front. Energy Res.*, vol. 11, July 2023, doi: 10.3389/fenrg.2023.1167043.

401f Leveraging Process Operability Mapping to Support Experimental Membrane Direct Air Capture (m-DAC) Solutions

Vitor Renan Vitor Gama, Oishi Sanyal, Fernando Lima

The urge for meaningful actions toward addressing legacy CO_2 emissions has led companies to reassess their current methods of operation, and researchers to consider a number of negative emissions technologies, including CO_2 direct air capture (DAC). Capturing CO_2 directly from the atmosphere, as first suggested by Lackner (2003)[1], requires technologies that are capable of extracting it from a very dilute source (CO_2 = 420 ppm). In this presentation, a multistage membrane-based direct air capture (m-DAC) system, based on novel facilitated transport membranes, will be described to produce low-purity CO_2 (~5% CO_2). The m-DAC process is modeled considering a hollow-fiber membrane module, since hollow-fibers provide high surface area to volume ratios, therefore leading to high productivities, which are critical for direct air capture applications. A comprehensive framework for a 2-staged membrane process, designed for this purpose, and recently developed in our group[2], is leveraged in this work. The design is simulated in AVEVA Process Simulation and utilizes the process operability framework through the Python package Opyrability[3], which helps to identify the effects of input parameters such as membrane intrinsic properties and surface areas on critical output parameters such as overall recovery, purity and energy consumption (kWh/kgCO_2). Aiming to determine DAC-specific membrane parameters, an inverse design study is conducted, based on previously acquired results[2], to pinpoint the desired intrinsic characteristic values for a membrane operating at DAC conditions. Furthermore, the assessment of a hybrid process combining adsorption with the proposed m-DAC design will be explored for enhanced performance, towards enabling membrane-based solutions for direct air capture.

References

[1] - Klaus S. Lackner, A Guide to CO2 Sequestration.Science, 1677-1678 (2003). DOI:10.1126/science.1079033

[2] - Gama, Vitor; Dantas, Beatriz; Sanyal, Oishi; Lima, Fernando V.; Process Operability Analysis of Membrane-Based Direct Air Capture for Low-Purity CO2 Production. ACS Eng. Au 2024; DOI: 10.1021/acsengineeringau.3c00069

[3] - Alves, V.; Dinh, S.; Kitchin, J. R.; Gazzaneo, V.; Carrasco, J. C.; Lima, F. V. Opyrability: A Python package for process operability analysis. Journal of Open Source Software 2024, DOI: 10.21105/joss.05966

401g Impacts of Environmental Conditions on the Stability of Aminopolymer Sorbents for Direct Air Capture

Yoseph Guta, Sichi Li, Marcos Andrade, Elwin Hunter Sellars, Amitesh Maiti, Simon H. Pang, Carsten Sievers, Christopher Jones

Amine-functionalized sorbents are primary candidates for large-scale CO_2 removal due to their unique chemistry that allows for the removal of ultra-dilute/trace CO_2 in high quantities. However, these sorbents lose their high CO_2 adsorption capacity when operating for extended adsorption/desorption cycling. Along with process upsets that induce undesired states (e.g. exposure to high temperature and high oxygen concentrations simultaneously), extended exposure to environmental components such as CO_2, O_2, and H_2O can contribute significantly to long-term sorbent instability.

This work employs advanced simulations and experimental studies to identify the role CO_2 and temperature play in accelerating oxidative degradation of a model aminopolymer sorbent (poly(ethylenimine)-alumina (PEI/Al_2O_3)). The results reveal that adsorbed CO_2 species (carbamic acid and carbamates) reduce the free energy barriers of C–N bond cleavage promoting chain branching reactions that enhance free radical concentration and further destabilize the sorbent. However, sorbed CO_2's tendency to accelerate oxidative degradation depends on several factors, including the presence of radicals on the chain, CO_2 loading, and temperature. For each CO_2-air mixture studied (0.04%, 1%, and 5% CO_2-air), there is a region of temperature (55 – 80 °C for 0.04% CO_2-air, 75 – 85 °C for 1% CO_2-air, and 80 – 100 °C for 5% CO_2-air) in which sorbent deactivation accelerates (more than 50% of the initial CO_2 adsorption capacity is lost) due to the combined effects of enhanced carbamic acid/carbamate-catalyzed C-N bond cleavage and increased mobility of the aminopolymers. Deep potential molecular dynamics simulations and experimental results show that increasing the CO_2 loading minimizes PEI branch mobility and increases the energy barrier for radical propagation, minimizing sorbent degradation. This work demonstrates the role CO_2 plays in sorbent stability and the importance of considering various atmospheric components in DAC sorbent design and development.

401h Hydrodynamics of a Pilot-Scale Dual Fluidized Bed Reactor: Cold Model Studies

Yongtie Cai, Preston Tan, Wen Liu

Calcium looping process is one of the cost-effective ways for post-combustion CO2 capture, utilizing CaO as a sorbent in the carbonator and $CaCO_3$ for CO_2 regeneration in the calciner. A pilot-scale cold model dual fluidized bed system, designed for this purpose, has been built at Waste-to-Energy Research Facility (WTERF) in Nanyang Technological University (NTU), consisting of a bubbling fluidized bed (BFB) as a carbonator for capturing CO_2 in the flue gas from WTERF and a circulating fluidized bed (CFB) as a calciner for regenerating CO_2 for further storage. The hydrodynamics of the pilot-scale system has been investigated through a series of cold model experiments. The effects of different operating conditions, including gas velocities in loop seals, carbonator and calciner, and solid inventories, on the pressure distribution and solid circulation rate have been studied. The preliminary objective of these cold model studies was to evaluate the feasibility of the design for the development of a hot model for continuous capture of CO2 from real flue gas. The results showed that the changes in the gas velocities resulted in variations in the pressure profile and axial solid holdup in the system. The gas velocities in the calciner played an important part in the total solid circulation rate. A stable operation was maintained throughout the system, determined by a pressure drop between the reactors. The implication of these studies contributes to the development of an efficient hot model for carbon capture and paves the way for the scale-up of calcium looping process for industrial applications.

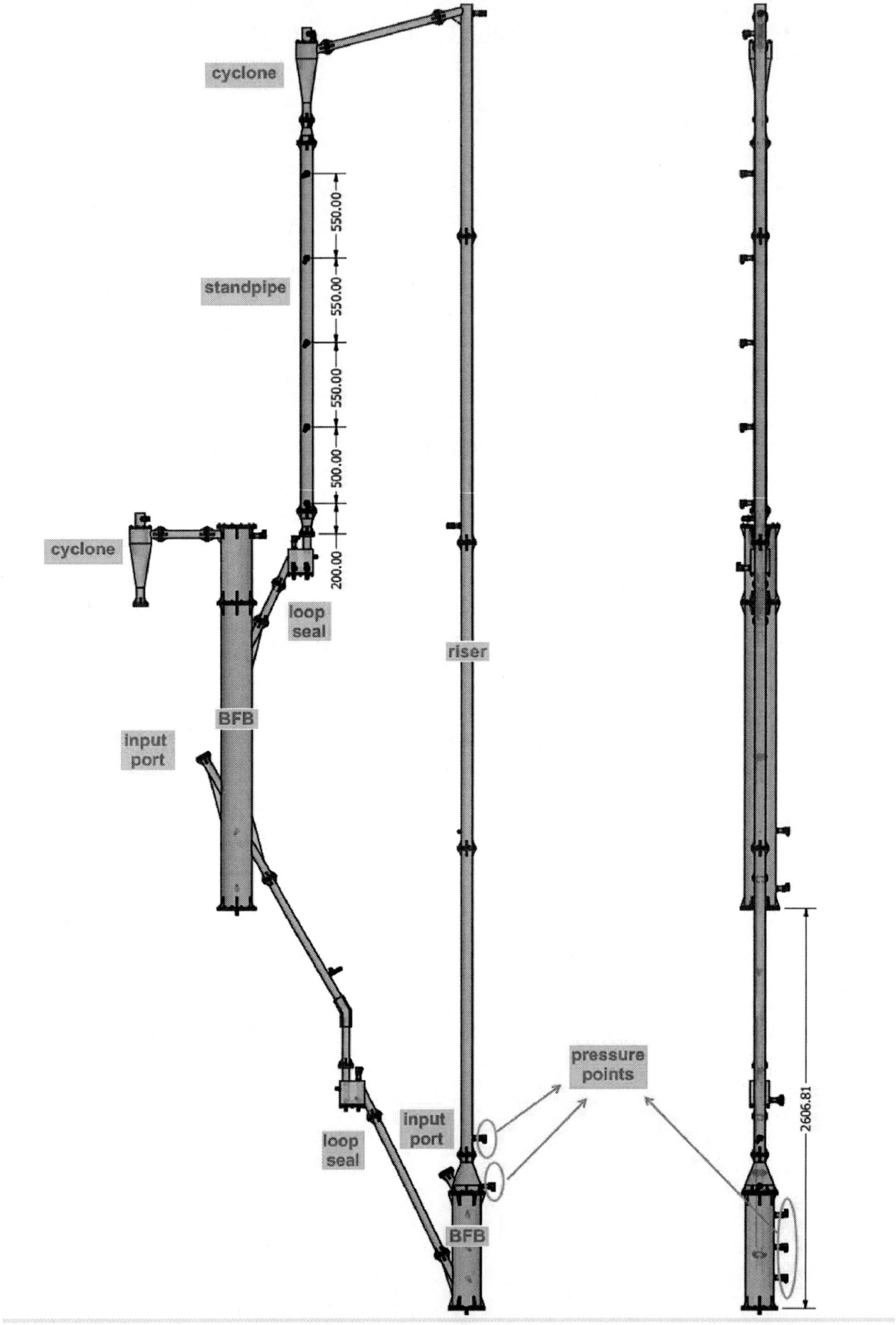
cyclone
standpipe
550.00
550.00
550.00
500.00
200.00
cyclone
loop seal
riser
BFB
input port
pressure points
input port
loop seal
BFB
2606.81

259b Techno-Economic Screening Analysis of Terrestrial Enhanced Weathering of Igneous Rocks and Industrial Waste Materials

Sarah Leptinsky, Sally Homsy, Hari Mantripragada, Timothy Fout

The Biden administration set the goal of achieving net-zero emissions by 2050. Meeting this goal requires investment in a portfolio of technologies including carbon dioxide removal (CDR). CDR technologies remove CO_2 from the atmosphere and can offer a means to decarbonize hard-to-abate CO_2 emissions (such as emissions from air travel) and meet the net-zero goal. Moreover, limiting global temperature increase to 1.5 °C in accordance with the Paris Agreement will require CDR to address legacy CO_2 emissions. One emerging CDR technology of interest is terrestrial enhanced weathering. Terrestrial enhanced weathering leverages the geologic process of weathering, which naturally removes CO_2 from the atmosphere. Weathering is the natural breakdown of alkaline rocks in the presence of rainwater, temperature changes, and/or living organisms. CO_2 in the air reacts with water to produce carbonic acid, which then reacts with alkaline rocks—containing hydroxide, silicate, and/or carbonates—to produce aqueous bicarbonate ions. The bicarbonate ions are eventually transported through soil, groundwater, and river water to the ocean where the CO_2 can remain in solution for more than 100,000 years. Sample reactions are presented in Figure 1.

While geologic weathering naturally removes CO_2 from the atmosphere, it is a slow process. Terrestrial enhanced weathering accelerates the geologic process of weathering by (1) crushing and grinding alkaline rocks and materials to increase their exposed surface area, and (2) distributing this material in favorable regions. Favorable regions include tropical areas or agricultural fields where the materials are exposed to water, microorganisms, and favorable pH and temperatures. The weathered material passively captures CO_2 over time and stores the reacted carbon in the form of a bicarbonate. Suitable materials for enhanced weathering include naturally occurring alkaline rocks or alkaline industrial waste. Examples of naturally occurring rocks include mined igneous rocks such as basalt (mafic rock) and dunite (ultramafic rock). Alternatively, alkaline materials other than rock can react with CO_2 in the atmosphere and, thus, be "weathered," such as industrial waste materials like biomass ash or cement kiln dust.

Terrestrial enhanced weathering has emerged as a technology of interest due to its simplicity, low energy requirement, and potentially low capital investment requirement. Due to the novelty of this technology, there is limited literature exploring the techno-economics of the process. This presentation reports on a screening techno-economic analysis that examines the impact of key factors that influence terrestrial enhanced weathering performance and cost.

Two cases are considered for analysis based on the material used:

(1) Naturally occurring mined igneous rock

(2) Industrial waste materials (like biomass ash or cement kiln dust)

The analysis considers locating the terrestrial enhanced weathering system in the Midwestern United States, because of proximity to alkaline materials, availability of farmland, and appropriate ambient conditions for weathering. For the igneous rock case, the size of the system is based on the average amount of igneous rock available from a single mine—250,000 tonnes/year. For the industrial waste case, the size of the system is based on the amount of suitable waste material produced by an industrial hub—150,000 tonnes/year.

The analysis scope for the igneous rock case includes material purchase, comminution, transport by truck, distribution of the material on farmland, and measurement, verification, and reporting (MVR) of the CDR. The comminution occurs at a remote location and is processed to a particle size of 20 μm. For the industrial waste case, the scope of analysis includes material purchase, transport by truck, distribution of the material on farmland, and MVR. Comminution is not required, since biomass ash/cement kiln dust is already available in small particle sizes. For each case, a performance model was developed to estimate the mass and energy balances of the process, and a cost model was developed to calculate the capital costs, operational and maintenance (O&M) costs, and the levelized cost of CO_2 capture (LCOC).

The cost of material transportation is estimated as a function of the amount of material and the transport distance. It is assumed that existing farming infrastructure can be leveraged to apply the material to the land; thus, application equipment is not included in the capital costs. MVR costs are estimated by assuming that soil samples are periodically evaluated in order to confirm CDR is taking place. The reacted material is reapplied each year, and a 30-year project lifetime is assumed. Capital costs include the equipment for comminution, material handling, and material storage. O&M costs include the material costs, transport costs, energy requirements, labor costs, and MVR. Capital and O&M costs are determined using in-house cost estimating references informed by engineering procurement and construction firm (EPC) input, engineering judgment, and literature-reported values.

Results from the techno-economic models show that the igneous rock-based process is able to capture about 140,000 tonnes of CO_2 each year while the industrial waste-based process can capture about 90,000 tonnes of CO_2 each year. The corresponding LCOC for the igneous rock case is about \$136/tonne of CO_2 captured. The industrial waste case results in an LCOC of about \$119/tonne of CO_2 captured. The resulting breakdown of the LCOC can be found in Figure 2.

Since many factors can greatly impact the performance and the cost of the terrestrial enhanced weathering, multiple sensitivity analyses were performed. Factors considered included the weathering rate and weathering potential, amount of material applied, material cost, site application, energy cost, MVR cost, and transportation distance. For the igneous rock case, a sensitivity on the rock size was also performed. Material size influences the material-specific surface area and comminution energy requirement based on the material size. For the industrial waste case, the specific surface area is varied.

The weathering rate and weathering potential were found to have a large impact on the LCOC for both cases (see Figures 3 and 4). This highlights the importance of material selection since the weathering rate and potential vary across materials of different compositions. Additionally, the weathering rate and potential are significantly impacted by ambient conditions; this highlights the importance of application location selection.

A preliminary life cycle analysis indicates the net CDR rate is >50% of the CO_2 capture rate (conservatively assumed). Therefore, the net cost of CDR is no more than double the reported LCOC. This screening analysis indicates that implementing terrestrial enhanced weathering by utilizing ultramafic rock and/or cement kiln dust in suitable locations can be relatively low cost.

This transparent screening analysis based on systematic techno-economic analysis provides preliminary insight into factors that impact the performance and cost of terrestrial enhanced weathering. The results indicate that terrestrial enhanced weathering has the potential to offer a feasible approach to CDR. Future work will leverage these results to design relevant case studies for a more detailed techno-economic analysis of terrestrial enhanced weathering.

Disclaimer

necessarily constitute or imply its endorsement, recommendation, or favoring by the United States Government or any agency thereof. The views and opinions of authors expressed herein do not necessarily state or reflect those of the United States Government or any agency thereof.

$$M(OH)_2 + 2CO_2 \rightarrow M^{2+} + 2HCO_3^-$$
$$MSiO_3 + 2CO_2 \rightarrow M^{2+} + SiO_2 + 2HCO_3^-$$
$$MCO_3 + H_2O + CO_2 \rightarrow M^{2+} + 2HCO_3^-$$

*where *M* is typically Magnesium (Mg) or Calcium (Ca)

Figure 1: Chemical weathering reactions

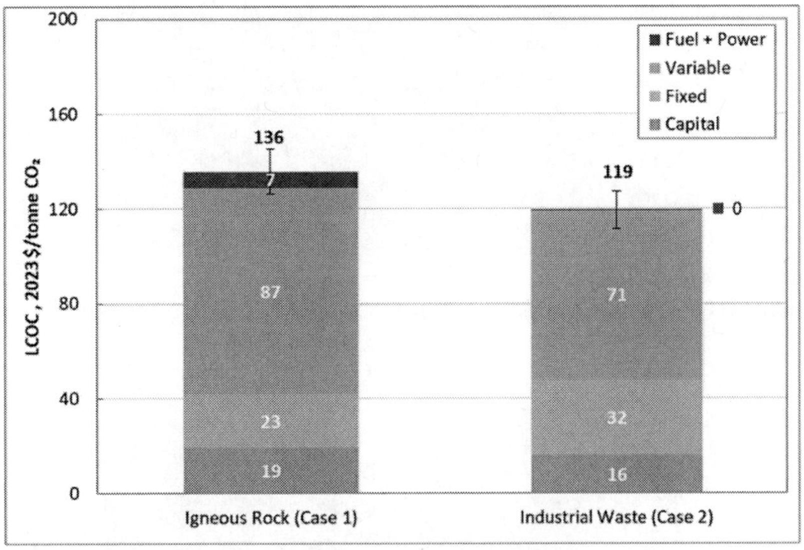

Figure 2: Levelized cost of capture

Weathering rate [mol·m⁻²s⁻¹]

(a) Weathering potential [kg CO_2/tonne]	1.E-08	1.E-09	1.E-10	1.E-11	1.E-12
200	--	--	--	--	--
300	316	316	362	1,673	14,759
400	237	237	272	1,255	11,069
500	189	189	217	1,004	8,855
600	158	158	181	837	7,380
700	135	135	155	717	6,325
800	118	118	136	627	5,535
900	105	105	121	558	4,920
1000	95	95	109	502	4,428
1100	86	86	99	456	4,025
1200	79	79	91	418	3,690
1,300	73	73	84	386	3,406

Mafic (300–800)

Ultramafic (1000–1,300)

Weathering rate [$mol \cdot m^{-2}s^{-1}$]

(a) Weathering potential [kg CO_2/tonne]	1.E-08	1.E-09	1.E-10	1.E-11	1.E-12	
200	--	--	--	--	--	
300	316	316	362	1,673	14,759	Mafic
400	237	237	272	1,255	11,069	Mafic
500	189	189	217	1,004	8,855	Mafic
600	158	158	181	837	7,380	Mafic
700	135	135	155	717	6,325	Mafic
800	118	118	136	627	5,535	Mafic
900	105	105	121	558	4,920	
1000	95	95	109	502	4,428	
1100	86	86	99	456	4,025	Ultramafic
1200	79	79	91	418	3,690	Ultramafic
1,300	73	73	84	386	3,406	Ultramafic

Figure 3: Terrestrial enhanced weathering with igneous rock LCOC sensitivity to weathering rate and weathering potential

Weathering rate [$mol \cdot m^{-2}s^{-1}$]

(a) Weathering potential [kg CO_2/tonne]	1.E-07	1.E-08	1.E-09	1.E-10	1.E-11	
200	358	358	358	534	3,503	Biomass ash
300	239	239	239	356	2,335	Biomass ash
400	179	179	179	267	1,751	Biomass ash
500	143	143	143	214	1,401	Biomass ash
600	119	119	119	178	1,168	Biomass ash
700	102	102	102	153	1,001	
800	90	90	90	133	876	
900	80	80	80	119	778	
1000	72	72	72	107	701	Cement kiln dust
1100	65	65	65	97	637	Cement kiln dust
1,200	60	60	60	89	584	Cement kiln dust
1,300	--	--	--	--	--	

Figure 4: Terrestrial enhanced weathering with industrial waste LCOC sensitivity to weathering rate and weathering potential

259c Comprehensive Sustainability Assessment of the Design of Achained Chemical Manufacturing System for Decarbonization

Abdurrafay Siddiqui, Yinlun Huang, Ying Liu, Qiang Xu

Decarbonization is the process of significantly reducing or eliminating GHG emissions including CO2. The chemical industry is the largest energy user among all industrial sectors. The GHG emissions footprint of the U.S. chemical industry was 274 million MT CO2 in 2020.[1] Thus, how to significantly improve energy efficiency and capture carbon in the chemical industry in a sustainable way is of great importance in industrial decarbonization. One approach for chemical industrial decarbonization is to achieve synergistic manufacturing among subsectors.[2] Qian et al. [3] and Liu et al. [4], among others, proposed conceptual design methodologies for constructing chemical complex systems to poly-generate chemical, fertilizer, power, and other products from natural gas and other resources, where GHG emissions were expected to be significantly reduced. However, the known design methodologies do not show whether the resulting complex systems could be highly sustainable.

In this presentation, we will introduce a comprehensive sustainability assessment method to evaluate a chained chemical manufacturing system designed by Liu et al. [5] for simultaneous production of ammonia, urea, liquified natural gas (LNG), and electricity, where internal carbon capture is realized. The assessment will be conducted using a sustainability metrics system that contains over 30 indicators, including one specifically quantifying decarbonization effectiveness. It is expected that the assessment will generate detailed information about whether such a type of chemical-energy-power integrated system design can demonstrate economic attractiveness, environmental cleanness (especially GHG emission reduction), and social responsibility clearly superior to the standalone design of the plants that are geographically distributed.

References

[1] U.S. Department of Energy, *Industrial Decarbonization Roadmap*, DOE/EE-2635, September 2022. https://www.energy.gov/industrial-technologies/doe-industrial-decarbonization-roadmap

[2] Zhang, K., H. H. Lou, and Y. L. Huang, 'Greenhouse Gas Emission Reduction via Advanced Heat Integration Techniques', in *Handbook of Climate Change Mitigation*, 3rd Edition, M. Lackner, B. Sajjadi, and W.Y. Chen, (eds.), Springer, 2021.

[3] Qian, Y., J. Liu, Z. Huang, A. Kraslawski, J. Cui, and Y. Huang, "Conceptual Design and System Analysis of a Poly-Generation System for Power and Olefin Production from Natural Gas," *Applied Energy*, 86(10), 2088-2095, 2009.

[4] Liu, Y.; Wang, S.; Xu, Q.; Ho, T. C. Eco-Friendly Natural Gas Monetization Complex for Simultaneous Power Generation and Nitrogen-Based Fertilizer Production. *Industrial & Engineering Chemistry Research*, 62 (1), 489-499, 2023.

[5] Liu, Y., S. Wang, Q. Xu, and Y. Huang, "Comprehensive Utilization of Natural Gas for Simultaneous Production of Power, Fertilizer, and LNG with Inherent Carbon Capture," submitted, *ACS Sustainable Chemistry and Engineering*, Sept. 2023.

The Cost Drivers of Solid-Based Direct Air Capture Technologies: Insights from Technoeconomic analysis

Sohaib Mohammed, Phuc-Tien Thierry, and Samuel Lethier

TotalEnergies E&P Research & Technology USA, TotalEnergies Tower, Houston, TX, USA 77002

Contacts: Sohaib.mohammed@external.totalenergies.com (SM); phuc-tien.thierry@totalenergies.com (PTT); samuel.lethier@totalenergies.com (SL)

Abstract

Atmospheric carbon dioxide (CO_2) removal via Direct Air Capture (DAC) has received increasing attention in the last decade. The International Energy Agency (IEA) predicts in its Net Zero Emissions by 2050 roadmap that DAC installations are projected to capture about 85 million metric tonnes (Mt) of atmospheric CO_2 by the year 2030, with anticipations of an increase to roughly 980 $MtCO_2$ by 2050. However, this scale up is faced with considerable challenges, mainly the elevated costs associated with the capture process. These costs are driven from the utilization of advanced sorbent materials and the energy intensive nature of the sorbent regeneration processes. Thus, further investigations on novel materials and process design are needed to reduce the capture cost and enable DAC deployment at scale.

Here, an engineering design of a solid-based direct air capture (S-DAC) unit and a technoeconomic analysis (TEA) model are developed to assess the cost of capture using S-DAC technology and enable performing sensitivity analysis on the primary cost drivers influencing the economics of S-DAC operations. This TEA model incorporates a wide range of input variables pertinent to adsorbent materials, contactor functionality in both adsorption and desorption phases, and the overall process design. Key variables such as adsorbent capture capacity, selectivity, lifetime, regeneration temperature, and contactor design are examined as input parameters to obtain the overall cost of CO_2 capture. The TEA findings highlight the significance of adsorbent materials as the primary cost drivers in S-DAC systems. It is observed that various properties of the adsorbent materials induce differing degrees of influence on the overall capture cost and can drive the cost up to > 2,500 $/tCO_2$. Moreover, it has been demonstrated that low-cost, durable and high-performing solid sorbent materials paired with contactor designs that facilitate high inlet air velocities alongside low pressure drops, has the potential to substantially reduce the cost of CO_2 capture.

The insights obtained from this work inform the significant role of the adsorbent materials performance and contactor design as substantial cost drivers of S-DAC technologies. These highlights are important to emphasize the need to develop novel and durable adsorbents and innovative contactor designs for efficient and economic S-DAC scale-up.

259e Techno-Economic Analysis of Water-Lean Solvent Carbon Capture Systems in Natural Gas Combined Cycle Power Plants for High Capture Rate

Shuang Xu, Yuan Jiang, Charles J. Freeman, David Heldebrant

According to U.S. Energy Information Administration (EIA), over 100 coal-fired plants were replaced or converted to natural gas between 2011 and 2019, while more conversions is expected in the future (EIA, 2020). Giving this trend, investigating low-cost carbon capture and sequestration (CCS) applicable to natural gas combined cycle (NGCC) power plants becomes important. However, efforts on carbon capture in NGCC power plants remained predominantly centered on aqueous amine solvent systems, which are usually energy intensive (Stowe & Hwang, 2017) due to the water vaporization/condensation within the process. Previous studies (Jiang et al., 2023) suggested that CO_2 Binding Organic Liquids (CO2BOLs), a class of water-lean solvent (with < 5wt% water) developed by Pacific Northwest National Laboratory, can achieve an estimated carbon capture cost below \$40/tonne CO_2, 15% lower than the industrial benchmark, for coal-fired power plant. The cost reduction was achieved by minimizing unnecessary water vaporization and condensation.

This study aims to investigate the economic viability of utilizing a water-lean solvent, specifically N-(2-ethoxyethyl)-3-morpholinopropan-1-amine (EEMPA), for NGCC power plant carbon capture. As emphasized by (Du et al., 2021), research on high capture rate from power plants is also important because negative-emission technologies such as direct air capture (DAC) are usually more expensive and less matured. This work further investigates the potential of employing water-lean solvent to achieve zero- and negative-emission for NGCC flue gas. Particularly, process model was developed in Aspen Plus V14, and techno-economic analysis was conducted for capture rates ranging from 90 to 99.8% (CO_2 concentration of treated gas <= 100 ppm). For each capture rate of interest, sensitivity studies were conducted to identify optimal operating conditions. For 90%, 95% and 97% capture, the estimated carbon capture cost of EEMPA was compared with NETL Baseline (Case B32B) (James III et al., 2022). For zero- or negative-emission regime, marginal cost, defined as the extra cost of capture additional amount of CO_2, was used to compare EEMPA with DAC technologies.

References

Du, Y., Gao, T., Rochelle, G. T., & Bhown, A. S. (2021). Zero-and negative-emissions fossil-fired power plants using CO2 capture by conventional aqueous amines. International Journal of Greenhouse Gas Control, 111, 103473.

EIA. (2020). More than 100 coal-fired plants have been replaced or converted to natural gas since 2011. U.S. Energy Information Administration. Retrieved Mar 22 from https://www.eia.gov/todayinenergy/detail.php?id=44636

James III, R. E., Shultz, T., Woods, M., Turner, M., Schmitt, T., Oaks, M., Konrade, J., Bleckinger, M., & Sturdivan, M. (2022). Cost and Performance Baseline for Fossil Energy Plants, Volume 3–Low Rank Coal and Natural Gas to Electricity.

Jiang, Y., Mathias, P. M., Zheng, R. F., Freeman, C. J., Barpaga, D., Malhotra, D., Koech, P. K., Zwoster, A., & Heldebrant, D. J. (2023). Energy-effective and low-cost carbon capture from point-sources enabled by water-lean solvents. Journal of Cleaner Production, 388, 135696.

Stowe, H. M., & Hwang, G. S. (2017). Fundamental understanding of CO2 capture and regeneration in aqueous amines from first-principles studies: recent progress and remaining challenges. Industrial & Engineering Chemistry Research, 56(24), 6887-6899.

259f Process Performance and Techno-Economic Analysis for Methane Mitigation from Low Concentration Sources

Devesh Sathya Sri Sairam Sirigina, Sai Gokul Subraveti, Simon Roussanaly, Rahul Anantharaman, Shareq Mohd Nazir

CH_4 is the biggest contributor to global warming among non-CO_2 GHGs [1]. The majority of CH_4 emissions are associated with low-concentration sources, which are notoriously hard to abate with agricultural sector being the largest source of anthropogenic CH_4 emissions [1, 2]. The technological options for methane mitigation via oxidation are present for confined sources of relatively high CH_4 concentrations (> 1000 ppmv) [1]. In addition to the existing preventive measures to limit CH_4 emissions, technological options to oxidize/convert them to a lower global warming potential gas can assist in mitigating the otherwise hard-to-abate or the already emitted gases. This is important in a larger context of achieving the net-zero GHG targets of respective sectors. The current work combines the mitigation of CH_4 followed by the capture of CO_2. The resulting process concept (s) can be applied to a multitude of low concentration CH_4 sources to reactively capture and mitigate the emitted greenhouse gas.

In our previous work [1], we presented process concepts to oxidize CH_4 followed by CO_2 capture for low concentration CH_4 sources. We presented two cases, namely, with CO_2 capture (co-removal of CH_4 and CO_2) and without CO_2 capture (CH_4 conversion), and analysed the effect of process parameters on energy penalty required to mitigate per ton of CO_2 equivalents in the system. The specific energy demand metric enables the comparison of these process concepts to direct air capture of CO_2 processes. The CH_4 oxidation process consists of a blower, a dehumidification unit, a recuperating heat exchanger, catalytic reactor, cooler, followed by CO_2 capture unit for the case with CO_2 capture [1]. CH_4 oxidation happens catalytically via thermal – and photocatalytic routes. A steady state model was developed and simulated in Aspen Plus to estimate the specific energy demand for CH_4 conversion. A model for CO_2 capture was not developed in this study but an energy estimate corresponding to a solid sorbent-based vacuum temperature swing adsorption (VTSA) process model from the literature [3] was considered. In the case with thermal catalytic route, a constant CH_4 conversion on 6.5 wt% Pd/Al_2O_3 [4] reported in literature is assumed. The amount of catalyst for the required conversion was later calculated using the kinetic model developed by Alyani et al. [4]. A similar process analysis for photocatalytic route using 0.8 wt% CuO/ZnO [5] as photocatalyst was performed. The results of energy demand by both the thermal- and photocatalytic routes were presented in [1].

Techno-economic assessment (TEA) for CH_4 oxidation for the thermal catalytic route yields a cost of 1258 \$/t-CO_2eq for the reference case with 300 ppmv of inlet CH_4 concentration in air. Figure 1 shows the results of TEA with different inlet CH_4 concentrations for CH_4 oxidation. Results for base case in CH_4 oxidation show that the recuperator and dehumidification unit dominate the capital costs with share of 41% and 45%, respectively. The costs of electricity and catalyst combined make up over two-thirds of the operating expenses. The costs presented in Figure 1 suggests a strong dependence on inlet CH_4 concentration, a result which is similar to energy demand (also shown in Figure 1). These results support a conclusion that for CH_4 oxidation to be economically feasible, either the catalysts which are insensitive to water vapor or higher inlet CH_4 concentrations are required. A higher CH_4 concentration can be achieved by including a pre-concentration step.

The current estimates for CO_2 removal through direct air capture were in the range of 200-600 \$/ton [6]. These estimates suggest that the co-removal (of CH_4 and CO_2) costs are likely to be lower than the cost estimate by just CH_4 oxidation (1258 \$/t-CO_2eq). In addition, a potential process heat integration between CH_4 oxidation and CO_2 capture unit was highlighted in [1] which in theory could further bring down the costs of co-removal. Alternatively, as noted before, the cost of CO_2 equivalents mitigation can also be lowered with higher inlet CH_4 concentrations. A CH_4 pre-concentration step (until 1 vol% CH_4) based on solid sorbent technology could potentially lower the overall cost of CO_2 equivalents mitigated. The TEA results for co-removal will be presented in the final work. The results will be compared to the process integrated with the pre-concentration step.

Sirigina et al. [1] reported the energy demand for inlet CH_4 concentrations varying from 100 ppmv to 1 vol% CH_4, and for CH_4 conversion in the range of 20%-100% in the reactor (Figure 4 in the reference [1]). The study however considers the catalytic oxidation of CH_4 to take place at 330 °C based on experimental results reported in literature for the 6.5 wt% Pd/Al_2O_3 [4]. The catalyst was reported to have a higher conversion at 380 °C [4]. In addition, a lower loss of catalytic activity by water vapor and thereby lower loss in CH_4 conversion in the experiments by adding Ce to 6.5 wt% Pd/Al_2O_3 catalysts is reported. As catalysts exhibit different activity with temperature (in addition to the effect of support, H_2O concentration etc,), and with temperature and conversion as proxy to capture the performance of a catalyst, a process performance map can be created reflecting the steady state activity of different catalysts. The process performance map in conjunction with a range of inlet concentrations representative of different emission sources could be used to identify the process performance of catalytic CH_4 oxidation on different catalysts. The available process model for CH_4 oxidation (which is largely independent of catalyst) will be used to analyse the effect of different temperatures on the performance in terms of specific energy demand per ton CO_2 equivalents mitigated. At higher temperatures (from the reference case at 330 °C), the waste heat from CH_4 oxidation will increase the potential for process heat integration between CH_4 oxidation and CO_2 capture. We will evaluate and compare the CH_4 oxidation and co-removal to identify the dynamic interplay at different temperatures. The performance map in terms of energy demand with temperature, concentration, and conversion will be presented in the final work. This will help in identifying an optimal operating region for a given catalyst system or be used to choose a catalyst system based on given inlet stream conditions. In addition, the final work also showcases the techno-economic potential of CH_4 oxidation, CH_4 oxidation with CO_2 capture (co-removal), and integrates both the pathways with a CH_4 enrichment step to assess their feasibility against the existing technologies.

Acknowledgements

The work is funded by:

(i) Energimyndigheten (Swedish Energy Agency) for the project "Energieffektiv negativa utsläpp från jordbrukssektorn" [In English: Energy efficient negative emissions from agriculture and farming] (project number 50340-1)

(ii) 'The Norwegian Continental Shelf: A Driver for Climate-Positive Norway' (NCS C+) project funded by the Research Council of Norway (328715) under the green platform program.

References

1. Sirigina, D.S.S.S., A. Goel, and S.M. Nazir, *Process concepts and analysis for co-removing methane and carbon dioxide from the atmosphere.* Scientific Reports, 2023. **13**(1): p. 17290.

2. Stolaroff, J.K., et al., *Review of Methane Mitigation Technologies with Application to Rapid Release of Methane from the Arctic.* Environmental Science & Technology, 2012. **46**(12): p. 6455-6469.

3. Sabatino, F., et al., *A comparative energy and costs assessment and optimization for direct air capture technologies.* Joule, 2021. **5**(8): p. 2047-2076.

4. Alyani, M. and K.J. Smith, *Kinetic Analysis of the Inhibition of CH4 Oxidation by H2O on PdO/Al2O3*

and CeO2/PdO/Al2O3 Catalysts. Industrial & Engineering Chemistry Research, 2016. **55**(30): p. 8309-8318.

5. Li, Z., X. Pan, and Z. Yi, *Photocatalytic oxidation of methane over CuO-decorated ZnO nanocatalysts*. Journal of Materials Chemistry A, 2019. **7**(2): p. 469-475.

6. Ozkan, M., et al., *Current status and pillars of direct air capture technologies*. iScience, 2022. **25**(4): p. 103990.

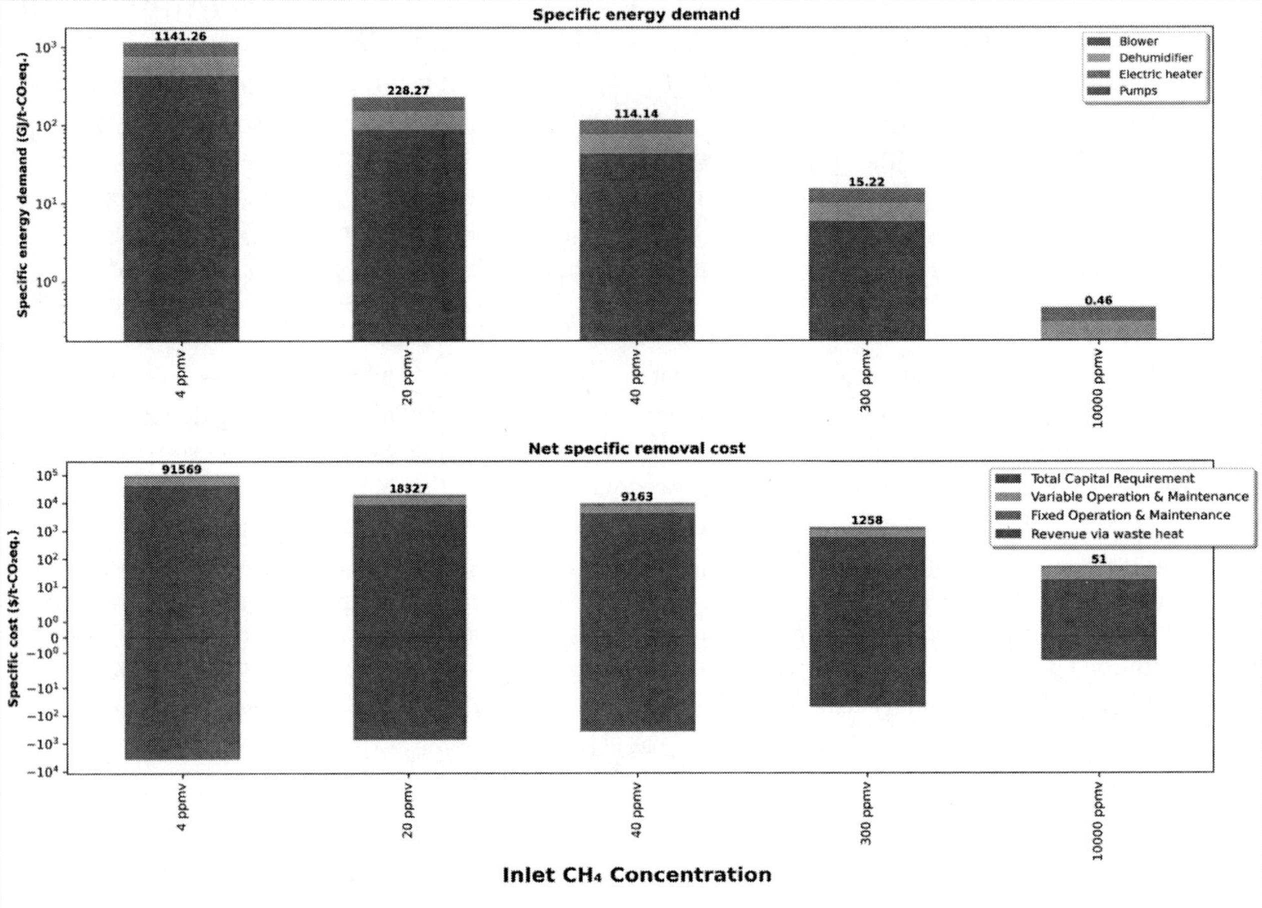

Figure 1: Process performance (specific energy demand) and TEA results of CH_4 oxidation for different inlet CH_4 concentrations.

259g Comprehensive Evaluation and Integration of Negative Emission Technologies

Paola Alejandra Saenz Cavazos, Nilay Shah

As the world grapples with the escalating impacts of climate change, the importance of Negative Emissions Technologies (NETs) has become increasingly evident. Amongst a diverse spectrum of strategies—from nature-based approaches like afforestation and reforestation, to technologically driven solutions such as Direct Air Capture (DAC) and Bioenergy with Carbon Capture and Storage (BECCS)—the challenge lies in identifying and optimizing the most effective methods. To address this, we developed a user-friendly scoring model based on different relevant Key Performance Indicators (KPIs).

The model uses qualitative and quantitative data obtained from peer-reviewed publications and expert interviews. A total of 14 different NETs were evaluated across more than 20 KPIs including technology readiness level (TRL), cost, removal capacity, energy and/or feedstock requirements, co-benefits, and trade-offs, among many others.

Figure 1 shows a snapshot of the model results comparing 8 different technologies across 5 critical KPIs. The model developed herein serves as a tool for delineating the strengths and limitations of each NET, as well as the synergistic potentials inherent in their integration into a cohesive portfolio. Model findings reveal that a strategically tailored combination of NETs, adapted to specific regional and environmental contexts, can significantly amplify carbon mitigation efforts. This integrative approach not only maximizes the effectiveness of each technology but also addresses broader ecological and socio-economic objectives. Moreover, the model facilitates the identification of collective knowledge gaps and emergent opportunities within NET portfolios, providing a roadmap for future research and development. The main identified collective areas of research include scalability, cost reduction, and the development of robust monitoring, reporting, and verification (MRV) frameworks.

This user-friendly model serves as a decision-support tool, outlining optimal integration strategies for various NETs, aiding policymakers, stakeholders, and engineers in crafting sustainable climate solutions and pinpointing further research opportunities.

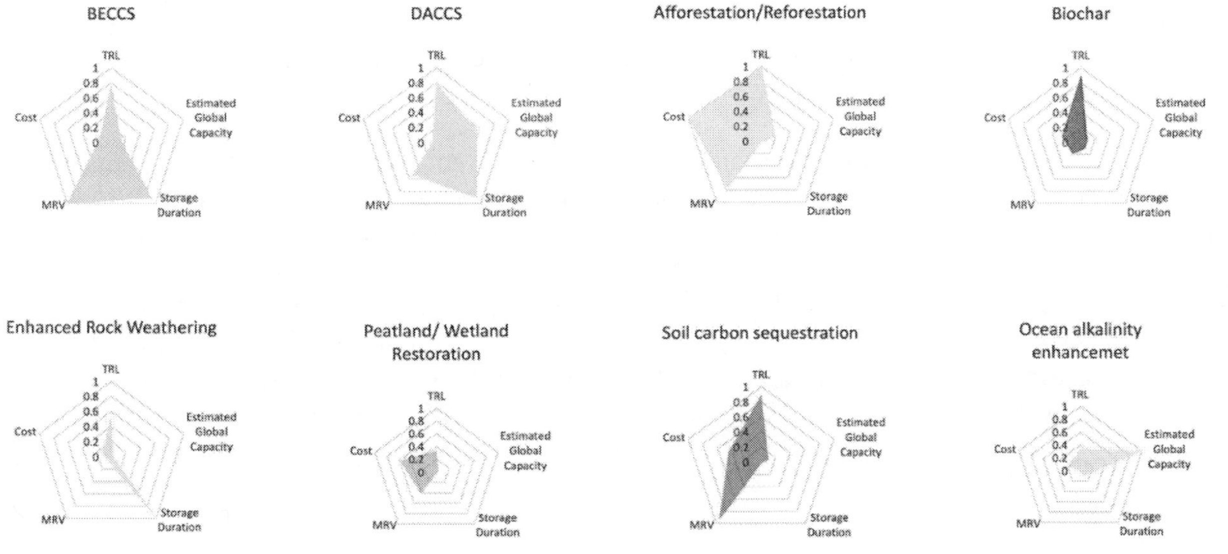

Figure 1. Model results comparing eight different Negative Emission Technologies (NETs) across five Key Performance Indicators (KPIs). Note that more extension towards Technology Readiness Level (TRL), Estimated Global Capacity, Storage Duration and Monitoring, Reporting and Verification (MRV) indicates higher parameter values while more extension towards cost indicates lower parameter values. Therefore, the larger area of the radar chart corresponds to a better overall performance of the technology.

260a Multi-Objective Optimization of an Integrated Cluster for Methanol Production

Sachin Jog, Daniel Vázquez, Ana I. Torres, Gonzalo Guillen-Gosalbez

Abstract

The Paris Agreement [1] has set a target of limiting the global average temperature rise to below 2 °C with reference to pre-industrial levels. With this aim, the chemical industry, which accounts for about 10% of global anthropogenic CO_2 emissions [2], is accelerating its efforts to transition from its reliance on fossil carbon to more sustainable processes. Further, to take advantage of the synergistic effects of integrating different feedstocks and production pathways, the design of chemical clusters has been gaining traction. This has resulted in extensive research on how to combine renewable carbon-based pathways, such as carbon capture and utilization (CCU), waste utilization, biomass utilization, etc. Additionally, there are ongoing efforts in the chemical industry to shift towards renewable energy sources (such as wind and solar power), which when combined with renewable carbon pathways can lead to a substantial reduction in environmental impacts in chemical clusters [3].

However, the intermittent nature of renewable energy sources remains a challenge when considering their use in the production of chemicals. Therefore, numerous studies assume certain simplifications to circumvent the dynamic modeling of renewable energy availability (*e.g.*, assuming an average capacity factor of the renewable energy source), which can result in overly optimistic conclusions [3]. To overcome these drawbacks, Martín and Grossmann [4] recently demonstrated the optimization of a process network for renewables-based fuels and power production, where the optimization was formulated as a multi-period problem with monthly time discretization. Further, Zhang *et al.* [5] showed the superstructure-based optimization of a process network for renewables-based fuels and power production, integrated with planning and scheduling. The multi-period problem involved an hourly time discretization, with operational constraints considered at the same level of granularity. Moreover, Corengia and Torres [6] developed a superstructure-based optimization framework with hourly time discretization for green hydrogen production, considering different renewable energy sources, different types of electrolyzers, and multiple storage options in the problem formulation.

In this work, we aim to integrate fossil and renewable routes into a chemical cluster and jointly optimize their economic and environmental performance considering an hourly time discretization of renewable energy availability. As such, the first step in the analysis is the modeling of the fossil and green methanol production processes. The fossil route consists of autothermal reforming (ATR) of natural gas for syngas production, which is subsequently used for fossil methanol production. On the other hand, the green route is based on methanol from the carbon dioxide hydrogenation process. Solid oxide electrolysis (SOE) powered by renewable energy is used to produce hydrogen (and oxygen) from water splitting, while direct air capture (DAC) provides carbon dioxide. The SOE is modeled using the Aspen Custom Modeler® v11, while all other processes are modeled in Aspen HYSYS® v11.

We consider two objective functions for the multi-objective optimization, *i.e.*, the annual production cost ($\$\cdot year^{-1}$) and the annual GWP impact (kg CO_2-eq$\cdot year^{-1}$). We formulate the multi-objective optimization as a multi-period MILP problem for a given methanol demand, also incorporating data on the hourly fluctuations of wind and solar power [7,8]. The problem is solved using the ε-constraint method implemented in GAMS v35.2.0 coupled with CPLEX v20.1.0.1 to find the optimal design and production schedule for methanol production, also incorporating storage of renewable energy, hydrogen, and methanol.

For a given maximum GWP impact target, our results provide the cost-optimal scheduling and planning of the production and/or storage of renewable energy, hydrogen, and methanol at each hour of operation over the course of one year. Further, the Pareto frontier obtained from the multi-objective optimization illustrates the hybridization potential between the fossil and green routes for methanol synthesis, which can enable a gradual transition to more sustainable production pathways in chemical clusters. Moreover, the comparison with the standard design method based on capacity factors showcases the advantages of the detailed multi-period model, particularly concerning the feasibility of the obtained solutions under dynamic availability profiles of renewables.

References

[1] United Nations, Framework Convention on Climate Change (UNFCCC), Adoption of the Paris Agreement (2015), (http://unfccc.int/%0Aresource/docs/2015/cop21/eng/l09r01.pdf), Accessed April 2024.

[2] F. Bauer, J.P. Tilsted, S. Pfister, C. Oberschelp, V. Kulionis, Mapping GHG emissions and prospects for renewable energy in the chemical industry, Curr. Opin. Chem. Eng. 39 (2023) 100881. https://doi.org/10.1016/j.coche.2022.100881.

[3] I. Ioannou, S.C. D'Angelo, Á. Galán-Martín, C. Pozo, J. Pérez-Ramírez, G. Guillén-Gosálbez, Process modelling and life cycle assessment coupled with experimental work to shape the future sustainable production of chemicals and fuels, React. Chem. Eng. 6 (2021) 1179–1194. https://doi.org/10.1039/D0RE00451K.

[4] M. Martín, I.E. Grossmann, Optimal integration of renewable based processes for fuels and power production: Spain case study, Appl. Energy. 213 (2018) 595–610. https://doi.org/10.1016/j.apenergy.2017.10.121.

[5] Q. Zhang, M. Martín, I.E. Grossmann, Integrated design and operation of renewables-based fuels and power production networks, Comput. Chem. Eng. 122 (2019) 80–92. https://doi.org/10.1016/j.compchemeng.2018.06.018.

[6] M. Corengia, A.I. Torres, Coupling time varying power sources to production of green-hydrogen: A superstructure based approach for technology selection and optimal design, Chem. Eng. Res. Des. 183 (2022) 235–249. https://doi.org/10.1016/j.cherd.2022.05.007.

[7] I. Staffell, S. Pfenninger, Using bias-corrected reanalysis to simulate current and future wind power output, Energy. 114 (2016) 1224–1239. https://doi.org/10.1016/j.energy.2016.08.068.

[8] S. Pfenninger, I. Staffell, Long-term patterns of European PV output using 30 years of validated hourly reanalysis and satellite data, Energy. 114 (2016) 1251–1265. https://doi.org/10.1016/j.energy.2016.08.060.

260b Analyzing Exergy Efficiency in Saint Louis City and Metro Area: A First Look

Kevin Noonan, Danahe Marmolejo, Clayton Stout

The rapid global urbanization mandates a meticulous examination of energy efficiency and availability in urban areas, forming the core focus of this proposal in Saint Louis City, Missouri and the Metro area. This study employs exergy analysis, deeply rooted in thermodynamics, as a key analytical tool. This approach allows us to evaluate both the quantity and quality of energy, providing a unique perspective beyond traditional assessments.

Building on the foundational works of Wall [1-2], Ertesvåg [3], and Hartmann & Garcia-Acevedo [4], this research aims to fill a critical gap by conducting the first exergy analysis specifically tailored to Saint Louis City. This deliberate choice of location enables us to unravel the distinct energy signatures of urban living, identifying areas with high energy availability and pinpointing inefficiencies in the energy landscape.

This interdisciplinary research is a work in progress, and we will be presenting the initial results of data gathering along with the first mapping of information. Sankey and exergy diagrams will illustrate the inputs and outputs of energy in the city, providing a visual representation of the energy flow and highlighting key contributors and consumers.

The approach presented here involves utilizing demographic data and quantifying energy and exergy inputs and outputs. Through the creation of diagrams and mapping of energy hotspots, we aim to actively engage in efficiency calculations, offering valuable insights into the energy dynamics of Saint Louis City.

The benefits of this research extend beyond academia. Understanding energy dynamics, particularly with an emphasis on exergy, contributes to informed political decisions and influences demographic distributions of energy and exergy. This work aims to contribute substantially to a nuanced comprehension of urban energy dynamics in Saint Louis City and the Metro area.

References:

1. Wall, G., 1981 revised 1986, Exergy Conversion in the Swedish Society, report no. 80-1, Physical Resource Theory Group, Göteborg. (Published in Resources and Energy, vol. 9, pp. 55-73, 1987.)
2. Wall, G., 1983 revised 1986, Exergy Flows in Industrial Processes, report no. 83-11, Physical Resource Theory Group, Göteborg. (Presented at 'the Fourth International Symposium on Second Law Analysis of Thermal Systems,' Rome, 25-29 May, 1987, I00236, pp. 131-140, ASME and published in Energy, vol. 13, no. 2, pp. 197-208, 1988)
3. Ertesvåg, I. S., 'Society exergy analysis: a comparison of different societies,' Energy, vol. 26, pp. 253-270, 2001. [Online].
4. Hartmann, R. M., & Garcia-Acevedo, L. E., 'Thermodynamic assessment of cities applying exergetic efficiency as an evaluation index,' Sustainable Energy Technologies and Assessments, vol. 50, 2022. [Online].

260c Data-Driven Robust Hydrogen Infrastructure Planning Towards Heat Decarbonization through a New Hybrid Decomposition Method

Xu Zhou, Margarita Efthymiadou, Lazaros G. Papageorgiou, Vassilis Charitopoulos

Given the increasing number of countries committing to a "Net-zero" emissions target by 2050, hydrogen as a low-carbon alternative of natural gas will become a significant player in advancing sustainable decarbonisation pathways [1]. Nowadays, there is a pressing necessity to develop a new hydrogen infrastructure network in the UK satisfying the increasing hydrogen demand to meet the ambitious target. Even though the hydrogen infrastructure planning problems on a nationwide scale have been widely explored in the literature [2,3], they ignore the demand uncertainty inherent to underlying problem, which could bring significant economic loss and even jeopardise the system security. Hence, the uncertainty-resilient schemes are required to alleviate these disadvantages. The common two ways to hedge against the uncertainty in the studies are stochastic programming (SP) [4] and robust optimization (RO) techniques [5]. However, often it can be challenging to get the probability distribution of uncertainties required in SP. On top of that, even if the stochastic process governing the uncertainty is fully or partially recoverable, the number of scenarios needed for its accurate representation tend to computationally intractable problems even for systems of moderate size. Thus, robust optimisation as an effective alternative offers a practical trade-off between feasibility and computational tractability.

In this work, we focus on the spatially-explicit hydrogen infrastructure planning problem with 5-year steps 2035-2050 and hourly resolution. This kind of complex energy-planning model typically involves representative days to alleviate the computational complexity considering their multi-scale nature. The issue of systematically accounting for the uncertainty introduced through the deployment of representative days within the planning models remains largely unexplored. To this end, we employ a data-driven adaptive robust mixed-integer linear programming optimization framework, where the demand uncertainty is captured through the introduction of uncertain representative days [6,7] and polyhedral uncertainty sets [8]. In the two-stage adaptive robust optimization (ARO, min-max-min problem), decisions are made on the "wait and see" basis, where the operational decisions like the hydrogen transmission flowrates across regions can be adjusted after the uncertainty realizations. This problem is NP-hard and cannot not be solved directly using off-the-shelf solvers. Therefore, we propose a new hybrid decomposition algorithm based on the column and constraint generation algorithm and block coordinate descent methods to handle it, which avoids the introduction of big-*M*linearizations [9], and can achieve a higher computing efficiency. Numerical results illustrate the effectiveness of the proposed framework and method with reductions in computational time exceeding 65% compared to monolithic approaches that are largely employed in the literature. Using as case study the decarbonisation of the heat sector in Great Britain, we further compare the two-stage ARO with the single-level static RO where decisions are made "here and now" without uncertainty feedback. The results show that ARO can lead to a lower system cost, and verify the advantages on controlling conservatism of ARO whilst allowing us to elucidate the added value of different flexibility options.

References

[1] R. Lowes, B. Woodman, Disruptive and uncertain: Policy makers' perceptions on uk heat decarbonisation, Energy policy 142 (2020) 111494.

[2] M. Moreno-Benito, P. Agnolucci, L. G. Papageorgiou, Towards a sustainable hydrogen economy:

Optimisation-based framework for hydrogen infrastructure development, Computers and Chemical Engineering 102 (2017) 110–127.

[3] S. K. Kamarudin, W. R. Daud, Z. Yaakub, Z. Misron, W. Anuar, N. N. Yusuf, Synthesis and optimization of future hydrogen energy infrastructure planning in Peninsular Malaysia, International Journal of Hydrogen Energy 34 (5) (2009) 2077–2088.

[4] J. R. Birge, F. Louveaux, Introduction to stochastic programming, Springer Science & Business Media, 2011.

[5] C. Shang, X. Huang, F. You, Data-driven robust optimization based on kernel learning, Computers & Chemical Engineering 106 (2017) 464–479.

[6] V. M. Charitopoulos, M. Fajardy, C. K. Chyong, D. M. Reiner, The impact of 100% electrification of domestic heat in great britain, Iscience 26 (11) (2023) 1–12.

[7] C. L. Lara, J. D. Siirola, I. E. Grossmann, Electric power infrastructure planning under uncertainty: stochastic dual dynamic integer programming (sddip) and parallelization scheme, Optimization and Engineering 21 (2020) 1243–1281.

[8] C. Ning, F. You, Data-driven decision making under uncertainty integrating robust optimization with principal component analysis and kernel smoothing methods, Computers and Chemical Engineering 112 (2018) 190–210.

[9] R. A. Jabr, Robust Transmission Network Expansion Planning With Uncertain Renewable Generation and Loads, IEEE Transactions on Power Systems 4 (28) (2013) 4558–4567.

Acknowledgements

Financial support under the EPSRC grant EP/T022930/1 is gratefully acknowledged by the authors.

260d Solar-Based Hydrogen Production and Storage System Utilizing Auto-Cascade and Mixed Refrigeration Cycles

Abbas Azarpour, Bahram Ghorbani, Sohrab Zendehboudi

Numerous countries based on the Paris Agreement have committed to significantly decrease carbon dioxide (CO_2) emissions and achieve net-zero carbon emissions by 2050. The signatories of the Paris Agreement are interested in increasing the use of hydrogen (H_2) as a clean fuel alongside other energy carriers. Large-scale storage and distribution of H_2 is a major challenge due to its low energy density in gas form. The main challenges of liquid H_2 storage as one of the most promising techniques for large-scale transportation and long-term storage include high specific energy consumption (SEC), low exergy efficiency, high total costs, and boil-off gas losses. In this study, an integrated structure of an auto-cascade and a multi-component refrigeration cycle is used for H_2 pre-cooling. Also, the Joule-Brayton process is employed to liquefy pre-cooled H_2. Parabolic trough collectors and photovoltaic panels are utilized to provide heat and electricity for the H_2 production and storage structure based on the energy potentials of Arkansas, USA. The proposed integrated structure utilizes a proton exchange membrane electrolysis electrolyzer for H_2 production. Utilizing the auto-cascade refrigeration cycle to recover solar energy in the form of refrigeration during the H_2 pre-cooling process results in a reduction of SEC. Pinch analysis is utilized to design heat exchanger networks effectively and reduce energy consumption. The exergy analysis of the proposed system provides valuable insights into the efficient use and loss of input exergy. Exergy loss, which acts as an irreversibility index, quantifies the inefficiencies within the system. In this research, the sensitivity analysis phase aims to investigate the effect of design variables on various output parameters, such as major energy factors (e.g., SEC and net power consumption) and exergy (e.g., exergy destruction and exergy efficiency).

260e Integrated Design and Operation for Micro-Electrochemical Plants Powered By HRES with Multi-Level Process Flexibility

Weigu Wen, Zhihong Yuan

The imperative target of carbon neutrality to tackle the global warming issue entails a rapid energy transition of the chemical industry towards renewable sources [1]. However, high penetration levels of renewable energy pose a staple challenge for chemical processes as the intermittent nature of renewable energy sources, predominantly solar and wind energy, conflicts with the long-term stable operation conditions of conventional chemical processes. It is gradually recognized that the flexibility of chemical operation units and processes plays significant roles in both the design and scheduling of chemical factories [2]. However, the methods for the integrated design and scheduling of flexible chemical processes with hybrid renewable energy systems (HRES), considering the flexible features including capacity adjustment, ramp rate, start-up and shut-down period, storage, and operation mode shift, are still developing [3]. Further investigations into the influence of system flexibility under fluctuating energy supply are also crucial to fully exploit these abilities. To address these problems, we formulated a multiscale optimization model for integrating the design and operation (scheduling, planning) of a flexible HRES-powered electrochemical synthesis system with multiple target products. The main electrochemical processes embody the electrochemical reduction of CO_2 (ECO_2RR) to CO followed by a Fisher-Tropsch synthesis, ECO_2RR to formic acid (FA) and electrosynthesis (ES) of H_2O_2. The electrochemical processes of the systems are outfitted with microreactors featuring different levels of flexibility. The optimization model thereby simultaneously reveals the optimal configuration of the system (the scales of the production units, energy output units, and corresponding auxiliary units) and preliminary operational policies based on the historical hourly weather data of representative weeks. Comparisons between the conventional production modes and flexible ones successfully indicate that, promoted by significant technological advances in electrosynthesis and microreactor design, chemical processes displaying sufficient flexibility are competitive and more compatible with HRES, with no large-scale energy storage systems and better tolerance of losses of power supply. Investigations on energy distribution demonstrate that energy efficiency and costs depend strongly on the complementary effects of different levels of flexibility. In general, more flexible processes consume more excess energy. However, the trade-off between flexibility and investment and O&M costs should be approximately made owing to the frequent operation adjustment. These results suggest the importance of process flexibility when electrifying the chemical industry in the near future.

References

1. Gailani A, Cooper S, Allen S, Pimm A, Taylor P, Gross R. Assessing the potential of decarbonization options for industrial sectors. *Joule*. 2024;8(3):576-603. doi:10.1016/j.joule.2024.01.007

2. Smith C, Torrente-Murciano L. The importance of dynamic operation and renewable energy source on the economic feasibility of green ammonia. *Joule*. 2024;8(1):157-174. doi:10.1016/j.joule.2023.12.002

3. Lazouski N, Limaye A, Bose A, Gala ML, Manthiram K, Mallapragada DS. Cost and Performance Targets for Fully Electrochemical Ammonia Production under Flexible Operation. *ACS Energy Lett.* 2022;7(8):2627-2633. doi:10.1021/acsenergylett.2c01197

260f Development of a PEMFC System Utilizing Metal Hydride and Hydrogen Peroxide for Unmanned Underwater Vehicle in Low-Oxygen Environments

Hyungjun Cheon, Dongmin Kim, Sinwoo Choi, Bongjae Lee, Heesook Roh, Joongmyeon Bae

In underwater applications such as Unmanned Underwater Vehicles (UUVs), the power system stands as a critical determinant of submerged travel distance. Fuel cell systems, boasting higher volumetric energy density compared to batteries, have gained significant attention as a power source for UUVs. This research focuses on the development of a power system for UUVs operating in low-oxygen environments, leveraging Polymer Electrode Membrane Fuel Cells (PEMFCs). The fuel cell system utilized metal hydride and hydrogen peroxide as hydrogen and oxygen sources, respectively. The metal hydride employed a AB2 type TiMn-based alloy as a hydrogen storage material. Hydrogen peroxide decomposed into water and oxygen through a catalytic reaction. The heat generated during the decomposition process was supplied to the hydrogen desoprtion in the metal hydride. The generated hydrogen and oxygen were supplied as fuel to the PEMFC and operate in dead-end mode to increase fuel utilization. This new concept fuel cell system, designed to operate in low-oxygen environments, demonstrated an average power output of 292.8 W for 92 hours.

127a An Exploration of the Mechanisms and Kinetics of Nanozymes from a Chemical Engineering Perspective.

Seckin Karagoz

Nanozymes are a category of nanomaterials that have inherent properties similar to enzymes. Currently, nanozymes are being used not only for molecular detection but also for environmental protection and cancer diagnosis and therapy. Nanozymes possess exceptional characteristics, combining the benefits of natural enzymes and artificial enzymes. These include remarkable activity and selectivity, adjustable activity levels, remarkable stability, and the ability to be reused. As a result, nanozymes have demonstrated significant potential and extensive applications in various fields such as medicine, agriculture, the environment, food, and pharmaceuticals. The potential uses of nanozymes create a new opportunity for nanomaterials. Nanozymes have enzymatic activities similar to oxidase, peroxidase, superoxide dismutase, and catalase enzymes. Nanozymes possess greater stability compared to natural enzymes and are capable of demonstrating activity across a broad spectrum of pH levels and temperatures. Nanozymes have lower production costs compared to natural enzymes, and they have the advantage of extended storage durations.

An essential inquiry in this domain is around the operational mechanism of these nanozymes. Most of the literature assumes that nanozymes exhibit enzymatic kinetics that are consistent with Michaelis-Menten kinetics. Nevertheless, there may be concerns over the overall applicability of this kinetic approach. Therefore, this study constructs and evaluates various kinetic models that rely on heterogeneous surface reaction processes using a comprehensive fitting algorithm. Ultimately, a comparative evaluation is conducted to analyze the Michaelis-Menten, Langmuir-Hinshelwood, and Eley-Rideal mechanisms and kinetics.

127b Environmental Pollution By Food Wastes and By-Products: Employing Probiotics for Their Valorized Conversion into Nutraceuticals.

Bababode Kehinde, Oluwabori Adekanye, Margaret Taiwo

In recent decades, food industries around the world have generated enough waste to significantly contribute to the net emissions of greenhouse gases and other forms of environmental pollution. Across the food supply chain: from cow to cream, sea to spoon and even farm to fork, large amounts of food waste are generated. On another note, the rise in demand for nutraceuticals and foods with enhanced health benefits has risen across the globe. This is partly associated with the rising cases of diverse health disorders that are supposedly associated with exposure to environmental pollutants arising from food processing factories. Several constraints have been observed to be associated with the cost-effective production of nutraceuticals. The prominent ones are related to raw material substrates, efficient processing methodologies, resulting impacts of such processes on the environment and the overhead costs involved. Probiotics have been known to impart desirable health benefits when consumed in adequate amounts. They are natural, sustainable (both in acquisition and usage), and biocompatible with the human system. Consumption of probiotics, food products processed with probiotics, or food materials with probiotics in their matrices have been observed to constitute minimal side effects relative to highly processed foods formulated with various additives. Food wastes have been studied and appraised as suitable ingredients that are useful as encapsulating carriers for the consumption and effective delivery of probiotics in the human gut. In addition, food macromolecules such as polysaccharides, lipids and proteins are usually degraded into smaller fractions of enhanced health benefits during probiotic fermentation. In addition, the employment of probiotics for both small and industrial scale processing of nutraceuticals portends a sustainable route with lesser deleterious impacts on the environment. The greener approach to processing nutraceuticals with probiotics will be discussed in this research session to further advance their application at all potential levels.

127c Nanoparticle-Impregnated Fibrous Adsorbent: A Novel Approach for Pb(II) Adsorption

Yashvi Sheth, Amritanshu Shriwastav, Rajdip Bandyopadhyaya

Rapid industrialization and urbanization have caused an increasing imbalance in the ecosystem because of the release of untreated or inefficiently treated wastewater streams in the water bodies. Amongst different pollutants, heavy metals like lead Pb(II) released from mining activities, electroplating industries, steel, and explosives manufacturing pose high toxicity. Pb(II) has even been detected in the groundwater and other water bodies with concentrations above its permissible limit for drinking water (10 ppb).

Currently, nanoparticle-based adsorbents have demonstrated promising and superior outcomes due to their high surface area-to-volume ratio and better selectivity towards heavy metals. However, they face challenges during scale-up attempts of the process, as their tiny size makes them susceptible to washing away, and their recovery is often tedious and impractical. Hence, the idea of this work is to synthesize a novel adsorbent by impregnation of nanoparticles onto a polymeric fibrous substrate in a manner that controls its leaching. The key advantage of using polymeric fibrous adsorbent is having a lower resistance to liquid flow (lower pressure drop) in fixed bed columns that would further minimize the channelling in contrast to conventional granular counterparts. Hence, the present work aims to apply this approach to the adsorption of Pb(II) from industrial wastewater to mitigate the repercussions caused by the discharge of Pb(II) containing water.

In this work, spherical-shaped hydroxyapatite nanoparticles (SHAP) having high adsorption capacity and selectivity for Pb(II) have been impregnated on nylon-6 fibres by two different in-situ modifications (i) Base treatment of fibres (resulting in SHAP-F), and (ii) $KMnO_4$ treatment of fibres (resulting in SHAP-M-F). The latter facilitated better loading of SHAP (5.1 wt%) as compared to the former (1 wt%) and subsequently led to MnO_2 nanoparticle formation on the fibre surface. This was confirmed from the SEM images as well as XPS analysis. From the SEM image in the figure, a uniform and dense distribution of nanoparticles on the surface of the fibre is observed.

Further, systematic batch adsorption experiments at varying concentrations of Pb(II) and pH were conducted. By fitting the batch results with isotherm models, the maximum adsorption capacities for SHAP, SHAP-F and SHAP-M-F were calculated to be 288 mg/g, 10 mg/g and 110 mg/g, respectively. These results highlight an enhancement in the maximum adsorption capacity by a factor of 3.63 and 4.73 for SHAP-F and SHAP-M-F, respectively as compared to the pristine SHAP. Additionally, it suggests a synergistic effect between MnO_2 particles and SHAP on the fibre surface, hence proving the potential of SHAP-M-F for Pb(II) removal.

Subsequently, to investigate the efficacy of industrial wastewater treatment, the batch process has been translated to a continuous lab-scale fixed bed process. SHAP-M-F adsorbent demonstrated the capability to treat water initially containing 10000 ppb of Pb(II) to a final concentration of 5 ppb in treated water which is well below the industrial discharge limit of 100 ppb, with a minimal residence time of 1 minute. Furthermore, the release of metal ions was monitored and found to be insignificant and below the discharge limits throughout the continuous operation.

Hence, from the present work, it can be inferred that nanoparticle impregnation on fibres offers a promising solution for leveraging their inherent properties and can effectively remediate Pb(II) from wastewater within a short contact time. This technique, can thus potentially be scaled up for applications towards industrial wastewater.

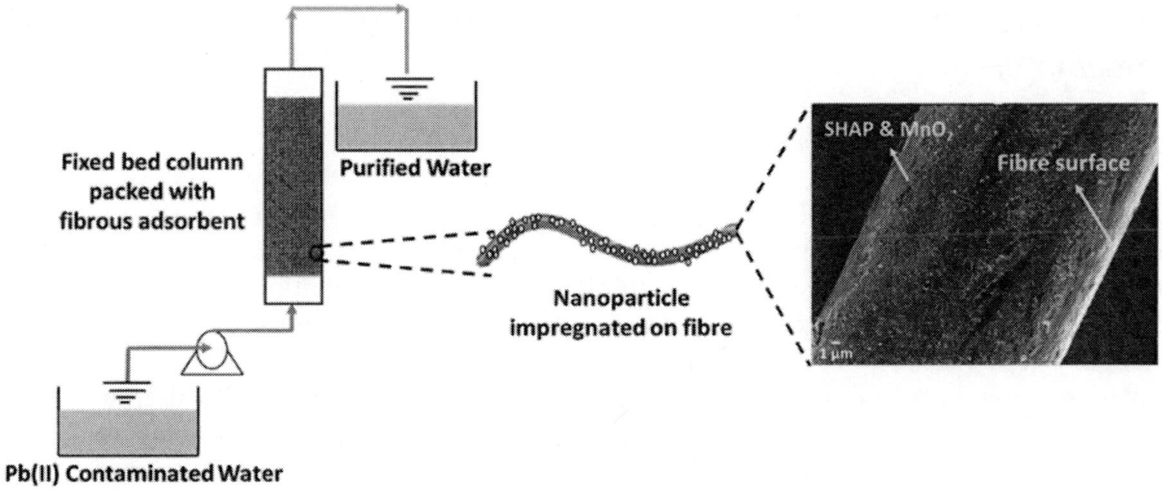

Figure: Schematics of the continuous Pb(II) adsorption on fibrous adsorbent

127d Leveraging Plant Nanobionics to Engineer the Next Generation of Phytoremediators

Cody Ritt, Gabriel Sanchez Velazquez, Michael S. Strano

Widespread environmental contamination by heavy metals, perfluorinated compounds, and a host of other toxic organics—paired with the rapidly progressing climate crisis—necessitates sustainable soil and groundwater remediation schemes. The use of plants for phytoremediation is a carbon-negative, self-sustaining, and inexpensive approach to remove and transform nonpoint sources of environmental contamination. Despite these attractive features, phytoremediation is exceedingly time consuming due to slow kinetics or low levels of biomass production, limiting their practical application to low-value lands. Herein, we quantitatively evaluate the potential for nanotechnology to enable the next generation of phytoremediators by conferring non-native functions to plants via introduced nanomaterials. We first model contaminant accumulation in plants as fluidic resistors-in-series, comparing common, hypertolerant, and hyperaccumulating plants to elucidate the bottleneck for rapid sequestration. This approach for contaminant transport is then extended to nanoparticle (NP) uptake itself to parameterize the predicted outcome of using nanotechnologies to enhance the rate of contaminant uptake in plants. We apply a thermodynamic lipid-NP interaction model to quantify design parameters for increasing NP root-to-shoot translocation in plants. Our findings suggest an order-of-magnitude increase in contaminant uptake rate is possible with NP-mediated schemes. Plant nanobionics thus lends itself as an exciting technique to engineer high biomass, non-hyperaccumulating plants as fit-for-purpose, next-generation phytoremediators.

127e Integrated Electro-Microbial Conversion of CO2 to Bioplastics

Peng Zhang, Kainan Chen, Bing Xu, Jinghao Li, Cheng Hu, Joshua Yuan, Susie Dai

Climate change due to greenhouse emission and plastics contamination are two significant environmental challenges for 21st century. A sustainable platform to convert CO2 to plastics is highly desirable as it presents a potential carbon neutral route to mitigate CO2 emission and plastics contamination. However, integrating catalytic CO2 reduction with bioconversion have been limited by inefficient electron and mass transfers, unfavorable metabolic kinetics, and inadequate molecular building blocks. We have overcome those challenges by designing a novel chemical-biological interface to integrate electrocatalysis and microbial fermentation. The soluble C2 intermediates produced from CO2 electrochemical reduction facilitate rapid mass transfer, present as favorable microbial substrates and less toxicity, and serve as an ideal molecular building blocks for longer carbon chain commodity chemicals. In this work, we present a multi-tier chem-bio interface design that has achieved six and eight times increase of microbial biomass productivity compared to the previous C1 intermediate and hydrogen-driven routes, respectively. The product, medium-chain-length polyhydroxyalkanoates (PHAs), as biodegradable polymers, presents a sustainable pathway to electrify the carbon tensive polymer industry and decarbonize the non-degradable plastics.

127f Modeling and Sensitivity Analysis of the Aqueous Oxidation Reactions of Nitric Oxide Induced By Thermal Activation of Peroxydisulfate

Yusuf Adewuyi

Pollutants emitted from combustion sources are of great public concern due to their detrimental effects on human health and ecosystems. Although multi-pollutants' reductions from flue gas using different processes such as selective catalytic reduction, lime/limestone wet flue gas desulfurization and activated carbon, respectively for NO_x, SO_2 and Hg removal, are efficient purification techniques, these complex treatment processes require high construction and operational costs, and large installation space. It is desirable to develop cost-effective especially those capable of providing simultaneous multi-pollutants' control. It is desirable to develop a new single integrated and cost-effective technology, which could also possibly be retrofitted to existing power plant. Advanced oxidation technologies (AOTs) are defined as those technologies that can generate mainly the hydroxyl radical (·OH) with high oxidation potential and other reactive oxygen species including superoxide anion radical ($O_2^{\cdot -}$), hydrogen peroxide (H_2O_2) and singlet oxygen, by various environmentally benign physical or chemical processes. Such technologies, which could also possibly be retrofitted to existing power plant are believed to be more cost-effective. Sulfate-radical () based advanced oxidation processes (AOPs) have drawn increasing attention in research and successful applications in water treatment and soil remediation as well as in flue gas purification for multipollutant treatments. The persulfate (PS) or peroxydisulfate anion ($S_2O_8^{2-}$) is a strong oxidant, kinetically slow at ordinary conditions but activation by heat, ultraviolet (UV), ultrasound, microwave, transition metal ions, electrodes, nanoparticles or base (and alkaline pH) resulting in the generation of sulfate (, and subsequent production of hydroxyl ($OH^{\bullet}$) radicals. Despite the recent numerous studies of NO removal by activated persulfate, the mechanisms are still not well understood. This work involves experimental studies and the development of a kinetic-mass transfer utilizing a comprehensive reaction scheme for process evaluation of the aqueous removal of nitric oxide (NO) in a simulated flue gas by heat-activated persulfate in a bubble-column reactor. For the first time, a sensitivity analysis and rate of production analysis are used to investigate the contributions of each homogeneous reaction to NO oxidation in the detailed mechanism to reveal the dominant pathway(s), and evaluate the effects on the predicted concentrations of key products. This talk will comprehensively evaluate the chemistry/kinetics, oxidant utilization (which impacts treatment cost), scale-up issues, and economic viability of these techniques for flue gas cleanup; and outline directions for the developments of innovative technologies for large-scale applications.

127g Carbon Ore-Derived Critical Materials for Clean Energy Technologies

Alexander Azenkeng

This presentation will focus on an emerging upgraded carbon ores to products (UCOP) technology for converting coal and coal wastes to high-quality graphite to assist in mitigating domestic critical graphite supply chains. Graphite has attractive applications for clean energy technologies such as active material for fabricating lithium-ion battery (LIB) electrodes, components for nuclear reactors, industrial electrodes, and lubricants as well as other uses in military applications. Trends in the manufacture of electric vehicles that use graphite-based LIBs are rising steeply in the United States and globally, but the U.S. domestic supply chains for graphite, the largest component in an LIB by mass, is severely limited and faces complicated geopolitical dynamics with foreign sources. Consequently, the United States has designated graphite as a critical mineral to focus attention and resources to develop technologies to meet the challenge of limited domestic graphite supply chains. Results obtained so far based on the UCOP process have successfully validated the technology at the laboratory scale, with the produced graphite material showing up to 95% degree of graphitization, high carbon purity of ~99.98%, residual ash content of ≤0.02%, negligible moisture, low trace elements, and high electrochemical stability. These results suggest that the emerging UCOP technology is a promising approach to effectively synthesize high-quality graphite from abundant coal and coal waste resources in the United States to create a sustainable domestic critical graphite supply chain. A brief description of the status of UCOP process development and representative results will be presented.

127i Investigating the Influence of PHA Polymer Coating Methods on the Biodegradability of Burlap Sacks for Shore Restoration

Roya Gadimli, Claire Thomas, Melissa Omand, Julie Albert

An important environmental problem that calls for creative restoration strategies is shoreline erosion. We investigate the effectiveness of coating burlap sacks containing glass sand with polyhydroxyalkanoate (PHA) polymers to restore shorelines. The aim of this study is to evaluate the effects of different coating techniques on the biodegradability of these bags in marine environments.

A number of PHA polymers are examined, such as poly(3-hydroxyvalerate) (PHV), poly(3-hydroxybutyrate) (PHB) and engineered commercial versions. To apply these polymers onto the burlap substrate and assess their efficiency in improving biodegradability, we studied several coating methods. The techniques we tried involve dip-coating, single-sided lamination and double-sided lamination. To identify structural changes brought about by the coating procedure, the crystallinity changes in coated burlap samples are characterized.

Degradation studies are also conducted to find out how coating techniques affect how long burlap sacks take to degrade in oceanic water settings. We hope to clarify the connection between coating technique, crystallinity variations, and degradation time by methodical testing and analysis, offering important information for the creation of sustainable solutions in coastal restoration projects.

65b Early Career Award Winner Presentation: Advancing Environmental Justice and Sustainability in Monitoring Emerging Air Pollutants

Marwa El-Sayed

Atmospheric particle pollutants, omnipresent in the atmosphere, pose significant risks to human health and influence climate systems. Although some aerosol types have been extensively studied, substantial knowledge gaps remain for other classes, leading to discrepancies between model predictions and real-world observations. This presentation aims to address these uncertainties by focusing on two lesser-understood classes of atmospheric aerosols.

The first focus is on secondary organic aerosols formed through aqueous processes. Our research seeks to characterize their formation and examine the effects of particle drying on their concentrations. To better understand these complex processes, we propose utilizing machine learning modeling techniques. The second focus is on nano- and microplastics, which present challenges in sampling, characterization, and identification. We explore these issues and propose innovative methods for their analysis.

To better characterize these pollutants and mitigate their harmful impacts, we employ low-cost sensor technology to map the high-resolution spatial distribution of aerosols, highlighting Environmental Justice (EJ) issues related to air quality. This approach provides insights into the equitable distribution of aerosol pollutants across different regions. By addressing these emerging pollutants, our goal is to close existing knowledge gaps and advance sustainability and EJ in atmospheric monitoring.

65a Cecil Award Presentation: Environmental Sustainability Analytics for Trending Issues on Virtual Engagement, Digital Assets, and Transportation Electrification

Fengqi You

In an era where energy and environmental sustainability are paramount, exemplified by the recent COP 28 Agreement and widespread "net zero" commitments, the role of sustainability analytics is increasingly significant. This presentation explores its applications in addressing key societal and industrial challenges. We will examine the climate impacts of recent shifts to virtual and hybrid conferences, the rising trend of remote and hybrid work models, and the metaverse industry's expansion. Subsequently, we will explore strategies for climate-neutral bitcoin mining and its potential to support renewable energy projects, along with the environmental aspects of non-fungible tokens and the Ethereum merge. The final part will discuss the critical metal requirements and environmental impacts of automotive batteries in transportation electrification, covering battery chemistry, recycling technologies, and global supply chain dynamics. Through these examples, we demonstrate the breadth of sustainability analytics, encompassing mathematical programming, data analytics, life cycle assessment, techno-economic analysis, integrated assessment models, systems design and integration, and model-based design of experiments, highlighting its critical role in advancing sustainability across various sectors.

716a Resolving Controversies in the Surface Area Measurements of Biochar and Hydrochar

Karen Agro, Kaivalya Gawande, Alex Maag, Geoffrey Tompsett, Wei Fan, Michael Timko

Waste biomass is an abundant and underutilized resource that results in greenhouse gas emissions from its decomposition and significant waste of valuable carbon when it has the potential to be thermochemically or biochemically converted into fuels or commodity chemicals. This waste biomass can serve as a sustainable feedstock to produce hydrochar and biochar materials. Hydrochar is produced through the thermochemical reaction of wet waste, without energy intense drying, at moderate temperatures (170 - 250°C) and autogenic pressures in a process coined hydrothermal carbonization (HTC). Biochar results from pyrolysis in the absence of oxygen to convert pre-dried wastes at high temperatures (400 - 800°C). Both types of char can be used as adsorbent materials, as a soil amendment, or as a biofuel. However, while hydrochar and biochar have clear similarities in their application and production method, their physicochemical structures differ significantly.

One key difference between hydrochar and biochar is surface area, which can be adjusted via post-synthesis activation. While biochar surface area has been more extensively characterized in literature [1], there is currently no optimized method for accurate measurement and analysis of hydrochar surface area. Since surface area is directly correlated with the material's adsorption capacity and performance in certain applications, it is paramount that hydrochar surface area is accurately measured and standardized. Furthermore, enhanced knowledge of hydrochar surface area will allow for a greater understanding of hydrochar synthesis and activation to enable tunable reaction engineering.

Currently, the most common method for measuring hydrochar surface area is through N_2 physisorption at low temperature (77 K) prior to analysis via the Brunauer–Emmett–Teller (BET) method. Traditional BET theory was developed for use on nonporous materials with a uniform surface and across a standard pressure range that does not account for the presence of micropores. The BET method's assumptions of a nonporous and uniform surface are inconsistent descriptors of the porous and functionalized surface of hydrochars. Therefore, hydrochar evaluated by N_2 physisorption often lead to unmeasurably small results inconsistent with the materials adsorptive performance in other media.

By utilizing the known similarities and differences between biochar and hydrochar, this study develops new methods to determine the surface area of hydrochar more accurately. Hydrochar was produced from glucose after HTC treatment at 180 °C for 12 hours in an autoclave reactor. A fraction of the hydrochar was pyrolyzed at varying temperatures ranging from 250 °C – 800 °C, to monitor the conversion to biochar. N_2 isotherm results on hydrochar output poor (negative) BET model fit C constants that indicate the resulting surface area are not physically meaningful. Due to the approximate size of hydrochar micropores (>2 nm) and the kinetic diameter of N2, this work compares the use of CO2 as an adsorbent gas in the determination of hydrochar surface area. However, the measured hydrochar surface area values were approximately 50 to 100 times larger when CO_2 was used as the adsorbate gas than N_2, with a valid model fit C constant. This can be seen in Table 1.

This study investigates hydrochar as well as pyrolyzed hydrochar/biochar, to develop and validate surface area methodology. This study develops an accurate procedure for measuring hydrochar surface area by utilizing CO_2 as the adsorbent gas at warmer temperatures (273 K), which has a higher quadrupole moment and smaller kinetic diameter than N_2, therefore interacting more with the abundant heteroatoms, and allowing for easier diffusion into the pores. This study also applies DFT to simulate the interactions between N_2 and CO_2 with hydrochar, thoroughly investigating both pore specific and pore non-specific surface area calculation theories and their physical meanings when applied to hydrochar. DFT aims to determine which combination of experimental methods and theory produces the most meaningful surface area values for hydrochar and evaluates the correlations between the methods and results. Our recommendations include using the Dubinin-Radushkevich theory and CO_2 at 273 K to measure a meaningful hydrochar surface area.

References:

(1)

Maziarka, P.; Wurzer, C.; Arauzo, P. J.; Dieguez-Alonso, A.; Mašek, O.; Ronsse, F. Do You BET on Routine? The Reliability of N2 Physisorption for the Quantitative Assessment of Biochar's Surface Area. *Chemical Engineering Journal* **2021**, *418*, 129234. https://doi.org/10.1016/j.cej.2021.129234.

Table 1. BET surface area measurements using N_2 and CO_2 physisorption

	N_2		CO_2	
Temperature (°C)	S_{BET} [m²/g]	C - constant	S_{BET} [m²/g]	C - constant
400	177.1	7.6	142.0	16.3
500	410.2	-52.6	205.7	13.8
600	430.8	-41.9	218.4	16.4
800	377.7	-44.1	277.4	21.9
180	1.0	-1.2	88.8	8.6

716b Systematic Investigation of Pore Structure Changes of Activated Biochar Using Small Angle X-Ray Scattering (SAXS)

Donovan Daubert, Hema Ramsurn

Conversion of biomass to biochar develops a complex porous network that has numerous applications in industry, such as carbon capture and environmental remediation. The pore structure and surface area are extremely important for controlling the effectiveness of biochar in all potential applications. This study is using small-angle x-ray scattering (SAXS) and electron microscopy to compare the pore size and shape distributions for activated biochar materials synthesized from model compounds (cellulose and lignin) and biomass from agricultural waste streams. The biomass will be converted to biochar using hydrothermal carbonization at 300 °C and 1300 psi, which are the optimal conditions to achieve high carbon content in biochar. The so-formed biochar is then thermally activated (in an inert nitrogen atmosphere) at various temperatures, ranging from 400-1000°C. Thermal activation induces re-organization of the amorphous carbon layers as well as the removal of oxygen-containing functional groups within the pore structure. Both of these processes result in nanoscale re-organization of the pores within the biochar (both size and morphology). Small angle X-ray scattering (SAXS) coupled with scanning electron microscopy (SEM) has been employed to monitor changes in the pore network and linked to the bulk characteristics of the biomass and activated biochar. We use these techniques to explore the hypothesis that the pore structure, pore evolution kinetics, and amounts/types of oxygen-containing groups in the activated biochar will be strongly affected by the activation temperature. Understanding the pore network development, especially the transition from micropore to mesopore, based on operating conditions can help guide the synthesis of the required activated biochar for specific applications.

716c Enhanced Carbon Capture on Functionalized Activated Hydrochar

Swarna Saha, Toufiq Reza

In recent years, CO2 emissions have emerged as a contentious issue due to their association with global warming and various other natural disasters. An economically viable adsorbent for CO2 capture could be produced from loblolly pine (LP), a common remnant of forests through functionalizing with carbon dioxide while performing hydrothermal carbonization. The objective of this study is to develop hydrochar from loblolly pine by pressurizing the reactor head while conducting hydrothermal carbonization at temperatures varying from 200°C to 260°C. The hydrochars are further super activated with KOH in a ratio of 2:1 at 800°C to synthesize a ultraporous adsorbent for comparing the effectiveness of CO2 capture. The activated characters displayed significant surface porosity and pore volume while characterizing. The adsorbent demonstrated improved surface functionalities and an increase in total oxygen content after functionalization with CO2. The functionalized activated chars demonstrated higher CO2 uptake at pressure range of 0.1 to 3 bar and 25°C. The enhanced CO2 capture efficiency was a result of the combined influence of increased surface functionalities and porosity, which improved the physical and chemical adsorption mechanisms.

716d Preparation and Characterization of Sulfonated Biochar and Its Application As a Photo-Fenton Catalyst in Water Treatment and Biomass Valorization

Goutham Rangarajan

Lignocellulosic biomass can be converted to biochar (BC), a material that possess excellent visible light absorption capacity and plenty of *sp*^{2}-hybridisation structures. In particular, BC-based catalysts provide access to innovative materials design owing to excellent surface properties, crystallinity, chemical stability, recoverability, and higher photocatalytic activities than pristine semiconductor photocatalysts. Recent studies providing evidence of resonance-stabilised free radicals (PFRs) on BC surface have highlighted the need to exploit the oxidative stress induced by PFRs for ecologically beneficial applications. Here, we report the synthesis and characterisation of H_2SO_4-modified BC-TiO_2 nanocomposite (SBC) as a solid acid to activate H_2O_2 at circumneutral pH under visible light radiation. In the prepared material, H_2SO_4 modification leads to removal of ash and improves surface area of BC, while TiO_2 acts as Lewis acid sites. We found under solar-simulating light that SBCs have a remarkable capability of activating H_2O_2 under Fe-free conditions, that can be exploited to drive oxidation reactions. Results showed that a complete MO removal and a TOC reduction of 70.2% can be achieved with SBC pyrolyzed at 500 ^oC. The catalytic activity of SBC depended on the pyrolysis temperature and correlated well with the surface acidity and persistent free radical (PFR) concentration. A combination of acid-site poisoning and EPR studies revealed that •OH is the key transient formed upon H_2O_2 activation, followed by $\bullet O_2^-$ and $^1O_2$. Our results demonstrate how a cheap, metal-free alternative heterogeneous catalyst could be developed from waste biomass to drive chemical reactions that are of industrial and environmental significance. We anticipate our carbon catalysts to be a starting point for more sophisticated materials design, leading to cheaper and robust nanomaterials, that can harvest light energy to drive oxidation reactions. For example, the activity of the SBCs could be expanded to biomass conversion, leading to a closed-loop photo-biorefinery for the valorisation of biomass into commercially valuable fine chemicals. Furthermore, this work offers a window into integrating biochar as a versatile, niche catalytic material in addition to its well known role as an adsorbent.

716e Solar Energy-Driven Biochar-Assisted Electrochemical H_2 Production

Rohit Chauhan, Nishithan Balaji Chidambara Kani, Meenesh R Singh

Hydrogen (H_2) is a promising source of clean energy since it has extreme heat value and eco-environmental advantages. The water electrolysis process is capable of producing high-grade H_2 and it requires 1.23 V cell potential to break water into hydrogen and oxygen molecules, but sluggish water oxidation kinetics result in a high loss of energy, which inhibits its application at the industrial scale. This sluggish kinetic can be improved by co-electrolysis with carbon-based sources such as agricultural wastes. Biochar derived from such wastes is known to drastically reduce the thermodynamic potential of electrolytic H_2 production from 1.23 V to 0.21 V by co-oxidizing biochar and water at the anode and generating protons for H_2 production at the cathode. Therefore, biochar advanced water electrolysis (BAWE) enables the high-value application of biomass with significant energy-saving production of hydrogen as a fuel. The application of BAWE to produce H_2 has several advantages: (i) biochar shows enhanced electrochemical activity than carbon black and graphite; (ii) it is applied at ambient pressure and temperature of >100 °C making it different from the conventional gasification process of biochar; (iii) the cathodic and anodic product i.e., H_2 and CO_2, respectively could be further utilized to produce methanol through cascade catalysis system; and (iv) there are several sources of biomass which could be utilized to synthesize biochar for BAWE like animal manure, bagasse, corn stover, etc. In this work, sub-volt H_2 production is demonstrated for the first-time using biochar derived from the chemical dehydration of five different agricultural wastes, such as sugarcane bagasse, paper waste, hemp waste, corn husk, and cow manure. The faradic efficiencies were calculated for the anodic and cathodic products, which were oxygenated hydrocarbons and hydrogen gas, respectively. Using a monocrystalline Si solar cell wired to a biochar-assisted water electrolyzer, a significant amount of current of ~15 mA of H_2 produced was obtained at 1 Sun. Biochar was characterized before and after electrolysis to compare the electrochemical reactivity, efficiency, and durability of the biochar.

716f Production of Biohydrogen and Valuable Bioproducts Via Anaerobic Fermentation with Engineered Biochars

Gyucheol Choi, Eunsung Kan

Organic waste treatment through anaerobic fermentation (AF) is one of the promising options for managing organic waste and recovering bioenergy sources. Biochar (BC), the carbonaceous material (CM) made from thermal conversion of various biomass is proposed as cost-effective and sustainable additive to enhance AF performance because of its low-cost feedstock and physicochemical properties such as specific functional group, high porosity, and high conductivity. In this study, a novel engineered carbon material (calcium-modified BC, Ca-BC) was test for enhancement of AF system along with commercial BC and activated carbon (AC). To explore the effects of each CM on AF, we examined biochemical acid potential (BAP) test with 5g/L of each CM and compared to condition without any additives (Control) under mesophilic and thermophilic condition. Under mesophilic condition, substrate utilization efficiency (SUE) reached 100% in 2 days at BC, followed by 3 days at Control and Ca-BC while AC showed slowest SUE achieving 100% in 4 days. It was interesting to see different final products with various CMs. BC showed the highest H_2 production while Ca-BC showed the lowest H_2 production although all runs achieved 100% of SUE. Meanwhile, under thermophilic condition, Ca-BC only achieved 100% of SUE during the experiment period (8 days), possibly due to its high buffering capacity. Interestingly, Ca-BC showed the lowest H_2 production even though it was the only run to achieve 100% of SUE, which is a similar observation with results under mesophilic condition. These results suggest that the surface properties of CM are crucial factors to determine the major metabolic pathway of fermentative bacteria.

716g Adsorption of Microcystins By Sargassum-Derived Biochars: Evaluation of the Characteristics, Mechanisms and Factors Influencing Adsorption.

Cadianne Chambers, Toufiq Reza

Sargassum, brown seaweed, accumulates in large quantities on ocean shores, causing visual and olfactory nuisances that affect tourism and fishery industries, with decomposing Sargassum producing hydrogen sulfide, prompting warning for individuals with respiratory illnesses to avoid affected areas. Thus, the valorization of this biomass to biochar presents as a sustainable option, as this material provides valuable surface qualities suitable for water purification. Due to climatic and anthropogenic events, increasing cyanobacterial blooms which release hepatotoxic Microcystins (MC) to the US freshwaters remain a concern livelihood of humans and aquatic life. Thus, the employment of Sargassum derived biochar for the remediation of these toxins, would be beneficiary for utilizing ocean's litter for the river's aid. Synthesis involved hydrothermal carbonization (HTC) at temperatures 180, 220, and 260 °C for 30 min followed by chemical activation with KOH at activation temperature 800°C with KOH to hydrochar ratio 2:1. Physical (Thermogravimetric Analysis, Ultimate Analysis), morphological (Brunauer-Emmet-Teller surface area, Scanning Electron Microscopy), and chemical characterization (Fourier Transform Infrared Spectroscopy, surface charge, Boehm Titration) were performed to study the resulting Seaweed Char based on the varying synthesis conditions. Batch adsorption experiments were conducted against Microcystins (Microcystin-LR and Microcystin-YR) for varying conditions (dosage amount, contact time, pH, and initial concentration). These investigations showed both physisorption and chemisorption mechanisms involving the uptake of Microcystins along with higher selectivity to Microcystin-LR. Consequently, use of Sargassum biochar has been shown to be highly efficient and cost effective for the purification of MC affected waters.

681a Hydrothermal Dechlorination of Municipal Solid Waste

Luke Walker, Aristidis Mihalos, Andrew Wagner

Plastics are a ubiquitous material in modern society due to their low cost, versatility, and durability, however traditional recycling is unable to meet the current demands of waste plastic disposal. As a result, as much as 85% of all waste plastic ends up in landfills. This is because most plastic waste is a mix of plastics and contaminated with food, dyes, and impurities. The expense to separate and decontaminate plastic waste is often prohibitive. Moreover, recycled plastic is often inferior to virgin plastic and therefore less valuable on the open market.

Thermochemical methods, like gasification, represent a feedstock-agnostic method to process mixed plastics and other waste to produce fuels and chemicals. However, halogenated plastics are a significant obstacle. It has been estimated that polyvinyl chloride (PVC) accounts for 12–15% of all plastic manufactured. Gasifying waste containing PVC releases chlorine gas that damages equipment and is a pollutant that has to be captured post-gasification. A pretreatment method that can remove chlorine prior to gasification has the potential to significantly improve gasification economics and reduce landfilling.

In this work, we examine the feasibility of using a hydrothermal treatment to remove chlorine from plastic in mixed waste, specifically PVC. Chlorine is captured as acid in subcritical water and the remaining solid materials are carbonized and highly amenable to gasification. We tested conditions to maximize chlorine removal and to retain energy in the solids for gasification. Products were analyzed to better understand hydrothermal dechlorination mechanisms and to evaluate the potential as a gasification feedstock. Finally, the results of the experiments were used in a technoeconomic analysis to determine the potential for treated mixed waste as an additional feedstock to improve gasification economics. We will discuss the effects of temperature, reaction time, solids to water ratio, and various catalysts on the treatment of pure PVC and mixed waste.

681b Site Location and Impacts of Valorization Operations

Robert Peters, Russell Fricano, Sandra Cutts

Valorization operations are not unlike recycling facilities. In contrast to landfills, location of valorization and recycling operations do not require extensive site analyses. Each operation can be located in an industrial zone. However, if the zone is within the vicinity of residential development, these facilities pose impacts including odor, truck traffic, noise and dust. This study examines the location of selected valorization plants near residential areas and examines potential impacts and whether the operation has a mitigation program.

681d Effect of Ultrasonic Pretreatment on Anaerobic Digestion to Remove Antibiotic Resistance Genes

Yasna Mortezaei, Goksel N. Demirer, Maggie R. Williams

Water resource recovery facilities (WRRFs) are crucial environmental sites where antibiotic resistance genes (ARGs) pose significant risks to human health. Anaerobic digestion (AD) is a common method used in WRRFs to manage organic waste and produce biogas, offering a potential means to mitigate ARG levels in sewage. Pretreatment strategies, such as ultrasonic (US) pretreatment, have been explored to enhance organic matter degradation and potentially improve ARG removal during AD. This study aimed to assess the impact of different durations of US pretreatment on the fate of ARGs in waste sludge. Our hypothesis was that US pretreatment could disrupt organic matter and microbial cells, thereby influencing ARG fate. Experiments were conducted using batch anaerobic reactors, wherein sludge from a WRRF in Mt Pleasant, Michigan, USA, underwent US pretreatment for varying durations (30, 40, and 60 minutes) before a 30-day digestion period. The relative abundance of ARGs (*sul1, sul2, tetW, tetA, tetO, ermF,* and *aac(6')-lb*) and mobile genetic elements (MGEs) (*int1, int2,* and *tnpA*) was assessed before and after US pretreatment, as well as after-AD, using quantitative polymerase chain reaction (qPCR). Results demonstrated a significant reduction in most detected ARG subtypes after US pretreatment, indicating potential disruption of cell structures and sludge integrity. Moreover, the relative abundance of studied MGEs showed a notable decrease after US pretreatment, suggesting limited horizontal gene transfer (HGT) events during this process. Subsequent AD further reduced the relative abundance of ARG subtypes, with the highest removal observed in samples subjected to US pretreatment for 45 minutes. Overall, these findings underscore the effectiveness of US pretreatment in reducing sulfonamide, tetracycline, macrolide, integrons, and transposase in sewage sludge, with optimal results achieved with a 45-minute US-AD treatment.

681e Navigating Environmental Justice in the Convergence of Water Valorization and Redevelopment: An Interdisciplinary Analysis

Sandra Cutts, Russell Fricano, Robert Peters

This presentation delves into the intricate interplay between water valorization and redevelopment within the context of environmental justice, shedding light on the complex dynamics that shape urban landscapes. Water, once relegated to a mere functional necessity, is increasingly recognized for its potential to drive economic growth and urban renewal. The process of valorizing water involves leveraging its diverse uses, ranging from recreational amenities to ecological restoration efforts, thereby positioning it as a linchpin in contemporary urban development strategies. Specifically, water valorization refers to the process of recognizing, enhancing, and maximizing the value of water resources through various means, including technological advancements, policy initiatives, and economic strategies. Rather than viewing water solely as a basic resource for human consumption, water valorization aims to leverage its multifaceted potential across environmental, social, and economic dimensions.

Concurrently, redevelopment emerges as a key mechanism for urban revitalization. By harnessing the potential of water resources as catalysts for renewal, redevelopment projects integrate water-centric approaches into comprehensive urban planning frameworks. However, these initiatives raise critical questions about environmental justice, as they may disproportionately impact marginalized communities through displacement, diminished access to resources, and heightened vulnerability to environmental hazards.

This presentation underscores the urgent need for an interdisciplinary approach to understanding and addressing the nexus of water valorization and redevelopment through the lens of environmental justice. By critically examining the complex interactions between these phenomena, policymakers, planners, and community stakeholders can develop strategies that prioritize equitable development, uphold community resilience, and foster inclusive urban environments that harness the transformative power of water resources for the collective benefit of all residents.

681f Hydrothermal Liquefaction of Municipal Sludge: Biochar, Biocrude, and Aqueous Co-Product Transformations with the Addition of Carbonate and Hydroxide Alkali Metals.

João Poli, Yanaidy Pagan, Daniel V. Tabaka, Susan Stagg-Williams

The hydrothermal liquefaction (HTL) of municipal sludge presents a promising avenue for sustainable waste management and biofuels production. Alkali metal catalysts have been shown to be promising for enhancing biocrude generation through waste-to-energy technologies by promoting depolymerization, deoxygenation, denitrogenation, and other reactions. Municipal sludge composition can vary substantially based on the types of treatment processes performed in a wastewater treatment plant. For example, some plants require supplemental alkalinity for nitrification which can impact the calcium concentration in the sludge and the resulting product distribution and characteristics. In this study, we have investigated the transformative effects of carbonate and hydroxide compounds on the three main HTL products: biochar, biocrude, and aqueous co-product (ACP). Municipal sludge was collected from the Wakarusa Wastewater Treatment Plant, located in Lawrence, KS. HTL was performed to determine the impact of the addition of calcium carbonate, calcium hydroxide, and sodium carbonate on the organic elemental composition, nutrient recovery, and product yields. Preliminary results suggest that the addition of Na_2CO_3 has the highest biocrude yield at 46.21 ± 1.23 (afdw%). The highest biochar yield was found to be 22.61 ± 0.64 (afdw%) with the addition of $Ca(OH)_2$. GC-MS results showed substantial change in biocrude composition depending on which chemical was added to the reaction. When no chemical was added, biocrude structure seems to have followed an acid pathway to mainly generate long chain alkanes and olefins. However, when carbonate and hydroxide alkali metals were added to the reactor, the reaction pathway changed towards the formation of amides. The addition of calcium carbonate and calcium hydroxide did not impact the CHNO composition of the biocrude with average values of 74.72% carbon, 9.7% hydrogen, 6.6% nitrogen, and 8.98% oxygen (by difference). However, the addition of sodium carbonate resulted in a lower biocrude carbon content of 71.97% and consequently higher concentration of oxygen in its composition at 13.6%. Further analysis of the biocrude properties, as well as the impact of the addition of calcium carbonate, calcium hydroxide, and sodium carbonate on the ACP and biochar properties, will also be discussed.

681g Turning Food Waste into Valuable Sorbents for Adsorptive Desulfurization

Henry Sokol, Julia A. Valla

Adsorptive desulfurization (ADS) has emerged as a promising alternative to conventional hydrodesulfurization, offering the prospect of producing sulfur-free fuels under ambient conditions with lower energy demands. Previous research has focused on ADS using zeolite sorbents with ion-exchanged cations to selectively adsorb high capacities of sulfur compounds from fuels.[1] Various surface metal species have demonstrated the ability to improve sulfur adsorption by providing available high-affinity active sites.[2,3] However, the available surface area of zeolites, in particular Y at 640 m^2/g, is decreased by metal ion-exchange and limits the maximum desulfurization potential. To overcome this challenge, higher surface area activated carbon (AC) materials have been considered as sorbents for ADS. While commercially available from many feedstocks including coal and biomass, AC derived from food waste (FWAC) has the added benefits of abundance and low cost while repurposing material otherwise destined for incineration or a landfill. FWAC materials have previously been successfully used as sorbents in other applications such as removal of phenol from wastewater.[4]

In this study we synthesized several FWAC materials via slow pyrolysis and subsequent physical activation by high temperature steam. By controlling the activation conditions, FWAC samples with a range of morphologies and characteristics were produced. These materials were characterized with various techniques to determine the physical and chemical properties and then tested for ADS using model fuels containing dibenzothiophene (DBT) and 4,6-dimethyldibenzothiophene (DMDBT) in octane with added aromatic hydrocarbons. The optimal FWAC sample with a surface area of 896 m^2/g was able to achieve maximum sulfur adsorption capacities of 6.3 mg S/g sorbent for DBT and 10.2 mg S/g sorbent for DMDBT. Characterization results indicate multiple mechanisms play a role is this excellent desulfurization. The synthesized FWAC samples show a linear relationship between micropore volume and sulfur capacity, suggesting physisorption of sulfur compounds in the micro pores of the carbon structure. In addition, ICP elemental analysis, XRD, FTIR, and TEM imaging show the presence of Ca, K, P, and Na species on FWAC. We hypothesize that these impurities from food waste provide suitable chemisorption sites for sulfur compounds. This is supported by results comparing FWAC to commercial AC and AC derived from other biomass sources including miscanthous, coconut shell, and walnut shell with similar surface areas. In these experiments the sulfur capacity of FWAC exceeded all other materials.

This application of FWAC represents a promising solution for reducing sulfur content in fuels, offering energy and cost savings compared to hydrodesulfurization. Moreover, the use of renewable materials as sorbents further minimizes the overall environmental impact of producing low sulfur fuels. The combination of impurities from food waste and tunable characteristics allow for high sulfur adsorption capacity exceeding other AC materials and zeolites. In addition, these attributes of FWAC may facilitate its use in various other adsorption and catalysis applications, transforming a waste resource into a valuable material.

(1) Lee, K. X.; Valla, J. A. Adsorptive Desulfurization of Liquid Hydrocarbons Using Zeolite-Based Sorbents: A Comprehensive Review. *React. Chem. Eng.* **2019**, *4* (8), 1357–1386. https://doi.org/10.1039/c9re00036d.

(2) Lee, K. X.; Tsilomelekis, G.; Valla, J. A. Removal of Benzothiophene and Dibenzothiophene from Hydrocarbon Fuels Using CuCe Mesoporous Y Zeolites in the Presence of Aromatics. *Appl. Catal. B Environ.* **2018**, *234*, 130–142. https://doi.org/10.1016/j.apcatb.2018.04.022.

(3) Lee, K. X.; Crowl, T. B.; Sokol, H. J.; Morales-acosta, M. D.; Valla, J. A. Understanding the Role of Rare Earths in Zeolite Y on the Removal of Sulfur from Hydrocarbon Fuels. *J. Phys. Chem. C* **2021**, *125* (17), 9107–9118. https://doi.org/10.1021/acs.jpcc.1c01103.

(4) Yu, L.; Gamliel, D. P.; Markunas, B.; Valla, J. A. A Promising Solution for Food Waste: Preparing

Activated Carbons for Phenol Removal from Water Streams. *ACS Omega* **2021**, *6* (13), 8870–8883. https://doi.org/10.1021/acsomega.0c06029.

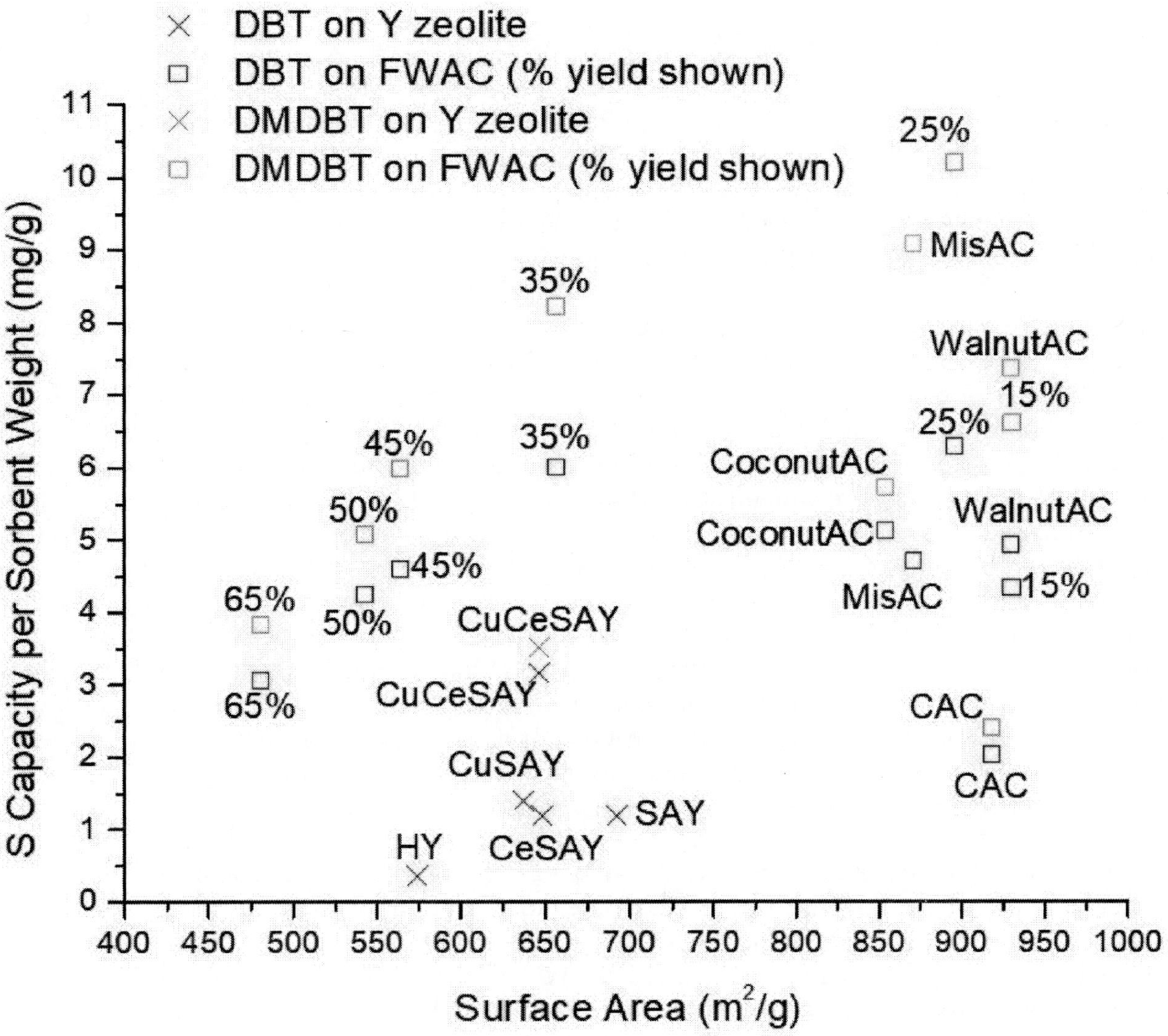

682a The Impact of Pollution from the Pesticide Atrazine, and Their Public Health Implications

Robert Peters, Gail Wallace, Christobel Asiedu, Ghislain Gueye, Teresa Murray

Atrazine is an herbicide that does not occur naturally. It is used to kill weeds and pests, primarily on farms, but has also been used on highway and railroad rights-of-way. Use of herbicides helps increase agricultural yields. Global pesticide usage is over 3.5 million tons in 2020; the largest users are the U.S., China, and Argentina. Other common herbicides used include glyphosate, 2,4-dichlorophenoxyacetic acid (2,4-D), metolachlor, and dieldrin.

Atrazine is used on a wide variety of agricultural crops as an herbicide to control broadleaf and grassy weeds. It can enter drinking water supplies from runoff from agricultural fields where it has been applied. Atrazine does not accumulate in living organisms such as algae, bacteria, clams, or fish. It can be taken up by plants and can accumulate in soils. Atrazine is hazardous to plants, soil organisms, and aquatic organisms. It is also a human health hazard; children and elderly people are particularly susceptible to health effects of atrazine. Consuming water with high levels of atrazine can cause adverse effects such as cardiovascular system problems, central nervous system disorders, and reproductive problems. Notably, there is no evidence that boiling water removes atrazine. Additionally, atrazine is an eye and skin irritant, and it may cause cancer. Liver, kidney, and heart damage have been observed in animals exposed to atrazine; we do not know whether this occurs in humans. Atrazine has been observed to cause changes in blood hormone levels that affect ovulation and the ability to reproduce. Exposure to high levels of atrazine causes a decrease in fetal growth and birth defects in animals, and causes reduced survival of fetuses in humans. A few studies have shown that it could affect pregnant women causing their babies to grow more slowly than normal or cause premature birth, and this has implications for brain health.

This presentation provides a description and review of various physical, chemical, and biological treatment methods for removing atrazine from water sources. Examples of treatment methods include: granular activated carbon adsorption; biochar adsorption onto zeolites and bentonite; Fenton advanced oxidation process; ozone oxidation, sulfate radical oxidation, photocatalytic treatment, microbial treatment, phytoremediation, plant-microbial remediation, and combinations of treatment technologies. Process performance will be indicated where available.

Health effects from atrazine exposure will also be addressed in the presentation.

682b A Holistic Approach to Biomass Thermochemical Treatment and Red Mud Recycling.

Pedro Henrique Cabral De Souza, Efthymios Kantarelis, Frederico M. Penha, Klas Engvall, Shareq Mohd Nazir

Problem:

It is estimated that over 4 billion tons of red mud are currently deposited in storage facilities worldwide, while 200 million tons of bauxite residue were produced globally in 2023 [1]. Red mud is the main by-product of the alumina refining process from bauxite ore, using sodium hydroxide for aluminum oxide extraction, producing a highly alkaline tailing[2]. Moreover, the bauxite residue can also exhibit radionuclidic activity [3]. Therefore, the dumping of red mud can cause severe environmental disruptions on disposal sites and surroundings, making it imperative to employ engineering structures to contain the residue. Thus, a new circular paradigm must be sought to tackle the red mud challenge. Rather than an unwanted residue, it is possible to face red mud as an important source of valuable materials. For instance, the major constituent of red mud, hematite, can be as high as 50 wt. % [4]. Other valuable metals, such as lanthanum, cerium, vanadium, and rare earth elements often add up to over 1000 ppm [5]. In addition, bauxite residue is also rich in oxides of aluminum, titanium, sodium, and calcium carbonate.

In this work, it is proposed the use of red mud as the oxidizing agent for biomass conversion, instead of air, carbon dioxide, or steam, as is customary. A gasification process that does not require the input of extra gases will yield a producer gas with higher HHV, given its undiluted nature. This will also result in a treated material in which iron is found in a reduced state, therefore recoverable by magnetic separation [6]. Given the high amounts of iron, alkali, and alkaline earth metal compounds in red mud, the bauxite residue is hypothesized to be catalytically active in both char conversion and tar reforming [7], [8]. This work aims to provide a holistic proof-of-concept of using red mud in biomass gasification with both gasification products and the treated red mud under evaluation.

Methods:

Bauxite residue samples used in this work were received from ETI Aluminum. Thermogravimetric experiments were performed to evaluate the mud-driven biomass conversion. Derived solid samples were collected and characterized by powder X-ray diffraction (XRD). Three samples were analyzed: 1) 10 ± 0.1 mg of pine sawdust; 2) 50 ± 0.5 mg of red mud, and 3) a mixture of 50 ± 0.5 mg of red mud with 10 ± 0.1 mg of pine sawdust. Thermogravimetric analyses were performed on a Netzsch STA 449 F3 apparatus. A 400 Nml/min nitrogen flow was employed as samples were heated from room temperature to 950 °C at a rate of 20 °C/min. A final 10-minute step was employed at 950 °C. Powder X-ray diffraction (XRD) was recorded with a PANalytical X'Pert PRO diffractometer with Cu Kα radiation ($\lambda = 1.54$ Å) at 40 kV/40 mA. samples were scanned from 5° to 80° (2θ) at a rate of 5° min^{-1} with 0.01° step size.

Furthermore, packed bed reactor experiments using red mud are carried out at similar conditions to TGA runs. Biomass and red mud samples are mixed at a mass ratio of 5:1 and exposed to 85 vol% CO2 (in N_2) gas stream. SPA method cand GC analysis is used for product characterization [9]. The resulting roasted red mud will be analyzed via XRD to ensure that the reduction behavior observed in TGA experiments is still valid when increasing the scale.

Results:

The thermogravimetric analysis results for the three samples are shown in Figure 1a in the form of DTG curves. Three mass loss events can be found for the red mud sample: drying of physically adsorbed water at low temperatures (<150 °C), decomposition of aluminum hydroxides into alumina is observed up to 350 °C [10]. The decomposition of carbonates is another important mass loss step observed in the characteristic 650 - 775 °C range, which is supported by the XRD spectra for the Red mud – 950 °C (Figure 1b) [11]. No change in the Iron oxidation state is recorded when red mud alone is taken to the TGA. The pine sawdust sample exhibits a typical lignocellulosic biomass behavior, with drying and pyrolysis. When pine and red mud were analyzed together, a synergistic effect was observed in three extra mass loss events at temperatures higher than 650 °C. This mass loss is due to the biomass-driven reduction of iron. This phenomenon is confirmed by the XRD spectra for red mud + pine – 950 °C, where no peaks for Iron III are found, but peaks for zero-valent Iron and FeO arise.

Implications:

This paper will present an experimental evaluation of red mud as an oxidizing agent in biomass conversion and a catalyst for tar reforming. Furthermore, the reduction of the iron content will also be assessed, given its importance for iron recovery. The proposed proof-of-concept establishes a disruptive synergetic framework for the simultaneous treatment of red mud and the valorization of waste biomass.

Acknowledgments:

The work is part of the project "Abtomat - Utilization Of Aluminium Bearing Raw Materials For The Production Of Aluminium Metal, Other Metals, And Compounds" (project number 2021-05201) funded by Vinnova (Sweden´s Innovation Agency) via the ERA-NET Cofund on Raw Materials (ERA-MIN3) program. The authors would like to thank the project consortium partners for the discussions. We would also like to thank ETI Aluminyum for the samples.

Bibliography

[1] M. Jovičević-Klug, I. R. Souza Filho, H. Springer, C. Adam, and D. Raabe, 'Green steel from red mud through climate-neutral hydrogen plasma reduction', *Nature*, vol. 625, no. 7996, pp. 703–709, Jan. 2024, doi: 10.1038/s41586-023-06901-z.

[2] E. Mukiza, L. Zhang, X. Liu, and N. Zhang, 'Utilization of red mud in road base and subgrade materials: A review', *Resour Conserv Recycl*, vol. 141, pp. 187–199, Feb. 2019, doi: 10.1016/j.resconrec.2018.10.031.

[3] B. Ozden, C. Brennan, and S. Landsberger, 'Environmental assessment of red mud by determining natural radionuclides using neutron activation analysis', *Environ Earth Sci*, vol. 78, no. 4, p. 114, Feb. 2019, doi: 10.1007/s12665-019-8120-8.

[4] S. Wang, H. Jin, Y. Deng, and Y. Xiao, 'Comprehensive utilization status of red mud in China: A critical review', *J Clean Prod*, vol. 289, p. 125136, Mar. 2021, doi: 10.1016/j.jclepro.2020.125136.

[5] C. R. Borra, B. Blanpain, Y. Pontikes, K. Binnemans, and T. Van Gerven, 'Recovery of Rare Earths and Other Valuable Metals From Bauxite Residue (Red Mud): A Review', *Journal of Sustainable Metallurgy*, vol. 2, no. 4, pp. 365–386, Dec. 2016, doi: 10.1007/s40831-016-0068-2.

[6] D. Zhu, T. Chun, J. Pan, and Z. He, 'Recovery of Iron From High-Iron Red Mud by Reduction Roasting With Adding Sodium Salt', *Journal of Iron and Steel Research International*, vol. 19, no. 8, pp. 1–5, Aug. 2012, doi: 10.1016/S1006-706X(12)60131-9.

[7] J. Yu, Q. Guo, Y. Gong, L. Ding, J. Wang, and G. Yu, 'A review of the effects of alkali and alkaline earth metal species on biomass gasification', *Fuel Processing Technology*, vol. 214, p. 106723, Apr. 2021, doi: 10.1016/j.fuproc.2021.106723.

[8] B. Ramadhani, T. Kivevele, J. H. Kihedu, and Y. A. C. Jande, 'Catalytic tar conversion and the prospective use of iron-based catalyst in the future development of biomass gasification: a review', *Biomass Convers Biorefin*, vol. 12, no. 4, pp. 1369–1392, Apr. 2022, doi: 10.1007/s13399-020-00814-x.

[9] V. Nemanova and K. Engvall, 'Tar Variability in the Producer Gas in a Bubbling Fluidized Bed Gasification System', *Energy & Fuels*, vol. 28, no. 12, pp. 7494–7500, Dec. 2014, doi:

10.1021/ef5015617.

[10] X. Yang, Z. Sun, D. Wang, and W. Forsling, 'Surface acid–base properties and hydration/dehydration mechanisms of aluminum (hydr)oxides', *J Colloid Interface Sci*, vol. 308, no. 2, pp. 395–404, Apr. 2007, doi: 10.1016/j.jcis.2006.12.023.

[11] N. H. Sulimai *et al.*, 'Synthesis of nanostructured calcite thin film by additive-free carbonation reaction via thermal chemical vapor deposition method', *Journal of Materials Science: Materials in Electronics*, vol. 32, no. 3, pp. 3072–3082, Feb. 2021, doi: 10.1007/s10854-020-05057-0.

Figure 1. a) DTG for red mud (50 mg), pine sawdust (10 mg), and a mix of red mud (50 mg) with pine sawdust (10 mg); and **b)** XRD spectra for raw red mud, red mud after the TGA, and red mud + pine sawdust after TGA.

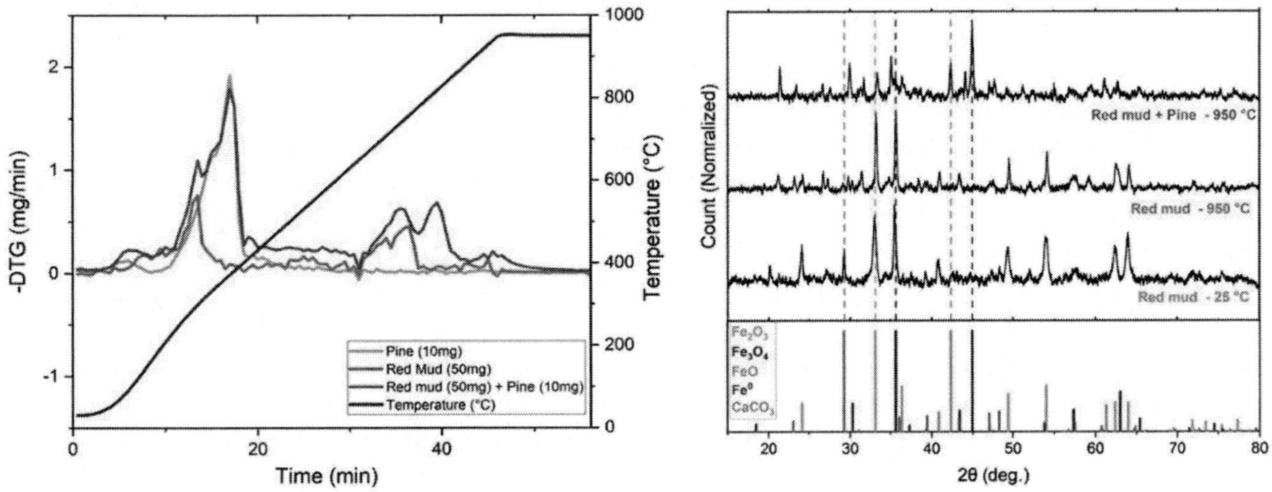

682c Advanced Light-Tolerant Microalgae-Nitrifying Bacteria Consortia for Stable Ammonia Removal Under Strong Light Irradiation Using Light-Shielding Hydrogel

Junichi Ida, Kiriko Aramoto, Beatriz Durán, Kento Nishi, Pabel Cervantes-Avilés, Germán C. Rodríguez, Shinichi Akizuki, Tatsushi Matsuyama

The consortium of microalgae and nitrifying bacteria has attracted attention owing to its advantages, such as energy- and cost-efficiency in terms of using only light irradiation without aeration. However, high light intensity can easily cause photoinhibition of nitrifying bacteria, resulting in process breakdown of the consortium. This challenge limits its practical application in outdoor environment. To overcome this challenge, we developed a "light-shielding hydrogel" which entrapped nitrifying bacteria in carbon black-added alginate hydrogel beads in this study and confirmed its effectiveness of protecting the nitrifying bacteria from intense light up to 1600 μmol photons m^{-2} s^{-1}. Then, we aimed to establish consortia of Chlorella sorokiniana and nitrifying bacteria immobilized in light-shielding hydrogel and evaluate their nitrification performance under strong light. Three nitrifying bacteria conditions were used: light-shielding hydrogel, hydrogel containing only nitrifying bacteria without carbon black ('hydrogel'), and dispersed nitrifier without immobilization ('dispersion') as a control. At 1600 μmol photons m^{-2} s^{-1}, the dispersion afforded a significant decrease in nitrification activity and subsequent process breakdown. In contrast, light-shielding hydrogel achieved complete nitrification without nitrite accumulation and had nitrification rates of approximately nine and two times higher than those for the dispersion and hydrogel conditions, respectively. Finally, the developed system was tested in Mexico using real anaerobic digestion effluent at the outside for long-term ammonia removal. In the 77-day experimental period, the average ammonia removal efficiency of 75.2% was obtained. This study demonstrated that the light-shielding hydrogel/consortia combination had potential for applications, which require mitigation of photoinhibition under strong light irradiation. Further, it is expected that the proposed method will contribute to realize the practical application of microalgae–nitrifying bacteria consortia in various countries that experience high sunlight intensity due to their location in the sunbelt areas.

WASTEWATER-DERIVED STRUVITE AS A GREEN ALTERNATIVE TO PORTLAND CEMENT: EXPERIMENTAL INSIGHTS INTO RHEOLOGICAL AND MECHANICAL PROPERTIES

Ugochukwu Ewuzie[1], Monday Okoronkwo[1] and Damilola Daramola[2], (1)Linda and Bipin Doshi Department of Chemical and Biochemical Engineering, Missouri University of Science and Technology, Rolla, MO, (2)Chemical Engineering, Northeastern University, Boston, MA.

Introduction

Globally, the construction industry is saddled with confronting responsibilities, including environmental impact, resource depletion, performance and durability, sustainability, and innovation. Cement, an inevitable binder (material) constituting the second most used commodity—concrete—has been implicated in contributing to greenhouse gas emissions [1]. With a sustainability mindset and the evolution of innovative engineering architectures, there is a continuous search for alternatives that extend beyond conventional supplementary cementitious materials (SCMs) [2]. These alternatives cannot only replace OPC and improve the characteristic performance of concrete but also contribute to contemporary environmental stewardship. Struvite (magnesium ammonium phosphate hexahydrate) is emerging as a dual-purpose SCM for replacing OPC.

Struvite precipitation from wastewater is an established and efficient phosphorus recovery technology to prevent eutrophication of water bodies [3]. In cement technology, struvite is the final and stable hydration product responsible for the high mechanical strength of magnesium ammonium phosphate cement (MAPC) [4,5]. Unlike in MAPC, only MgO is added to precipitate ST since phosphorus is already present in wastewater in sufficient amounts. Notably, Mg^{2+} (from MgO) is required because ST precipitates at a 1:1:1 molar ratio for Mg:N:P when the precursors' molar concentration exceeds ST's solubility product constant [6]. The dead-burned MgO used in conventional ST precipitation is derived from natural magnesite treated to high temperatures, between 1500 and 1800 °C [19], which is energy-intensive, emits enormous CO_2 and is a depleting natural resource. However, current low-cost MgO sources make ST generation from wastewater more sustainable and economically feasible since MgO contributes up to 75% of the overall cost [7].

Wastewater treatment by-products may be added to the waste streams explored in the construction industry to replace OPC sustainably. However, studies into the valorization of wastewater-generated by-products to benefit infrastructural development are scarce. Besides mechanical strength, the study of the rheological behavior of cement pastes, including yield stress and viscosity, is vital to predicting the workability of the paste. Studies that provide foundational insights into the mechanical, rheological, and microstructural behavior of ST materials in cement are scarce. Therefore, this pioneering study investigated the substitution of OPC with struvite, which can be produced from wastewater. In addition, it provided insights into 1) the hydration kinetics of CST pastes, 2) the effect of increasing ST content on the rheological behavior, and 3) the effect of increasing ST content on the porosity and mechanical strength of the CST blend. This study will serve as a foundational study for the potential use of CST blend in concrete technology, especially 3D printing.

Methodology

The Portland cement (OPC: type I) was sourced from Continental Cement Co., MO, USA. Reagent-grade struvite was acquired from ThermoFisher Scientific, USA. The elemental oxide composition of ST and OPC, as analyzed by X-Supreme 8000 x-ray fluorescence (Oxford instruments, UK), is detailed in Table 1. The zeta potential and particle size distribution of the cement and ST samples were assessed using the Litesizer 500 Dynamic Light Scattering (DLS, Anton Paar, GmbH) instrument. Also, the struvite crystals and OPC and their hydration products were examined using the X'Pert X-ray diffraction equipment (PANalytical MRD, UK). The compressive strength test instrument was a Tinius Olsen universal compression instrument (Tinus Olsen Inc., PA, USA), while the flexural strength test instrument was an Instron Tensile Compression Test Machine (Instron, MA, USA.). The hydration kinetics was measured with the THT μRC single-channel isothermal micro reaction calorimeter equipped with Peltier-based temperature control with resolutions ranging between 5 μW and 600 mW. A Discovery Hybrid Rheometer (DHR-2, TA Instruments, U.S.A) with a crosshatched 40 mm parallel Peltier plate geometry was employed to measure CST pastes' deformation and flow properties. Zeiss Xradia 620 Versa Micro X-ray computed tomography (μXCT) and 3D image analysis using Dragonfly version 21 were employed for porosity assessment. The mix design is presented in Table 1.

Table 1. Mix proportion and paste flow behavior

Notation	Cement (%)	Struvite (%)	Plastic viscosity (Pa.s) (fixed w/b = 0.50)	Dynamic yield stress (Pa) (fixed w/b = 0.50)	Static yield stress (Pa)	w/c†	Plastic viscosity (Pa.s) (fixed w/c = 0.50)	Dynamic yield stress (Pa) (fixed w/c = 0.50)	Static yield stress (Pa)	w/b‡
CST0	100	0	0.67	20.9	43.6	0.50	0.67	20.9	43.6	0.50
CST3	97	3	0.39	20.2	32.9	0.52	0.56	30.0	41.6	0.49
CST10	90	10	0.27	16.6	22.8	0.56	0.45	49.3	65.8	0.45
CST15	85	15	0.26	21.7	25.4	0.59	0.46	65.7	110.3	0.43
CST20	80	20	0.06	1.54	21.0	0.63	0.53	126.6	171.6	0.40

†w/c = water-to-cement ratio; ‡w/b = water-to-binder (total solid) ratio. The mix design with 0.5 w/c was adopted for subsequent rheological and mechanical strength assessments. The number in front of CST indicates the percentage of ST in the mix (CST0 is the control).

Figure 1 represents the mixing and measurement of CST samples. The same mixing procedure was employed for mechanical and rheological assessments. In both assessments, a maximum of five-minute interval between adding water to the cement and the rheological testing or curing was allowed. Important rheological properties of CST paste, which provide critical information about its behavior in the fresh state, are elucidated in Fig. 2. In Fig. 2a, the peak shear stress where the flow commenced indicated the static yield stress obtained at a very low constant shear strain rate (0.01 s^{-1}). Besides static yield stress, dynamic yield stress and plastic viscosity can be derived by fitting the downward flow curve with a modified Bingham model, as shown in Fig 2b. More importantly, the evolution of static or dynamic yield stress as a function of hydration time (Fig. 2c) enables the estimation of workability metrics, such as CST paste flowability loss rate, hardening rate, and placement limit. For CST pastes porosity tests, 10% struvite substituted cement paste (CST10) was used and tested after 1 and 28 curing days.

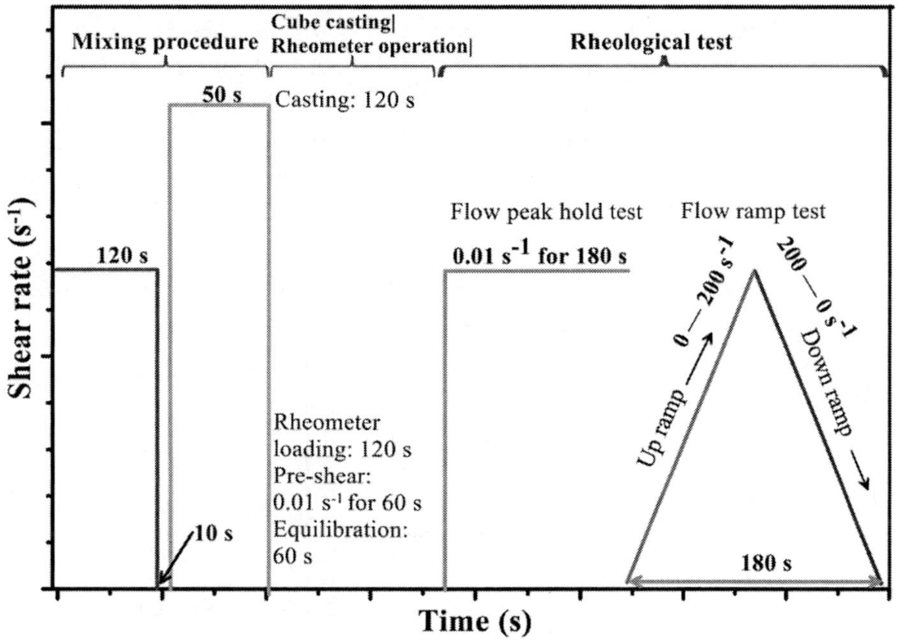

Figure 1. Mixing and measurement protocol for the CST pastes

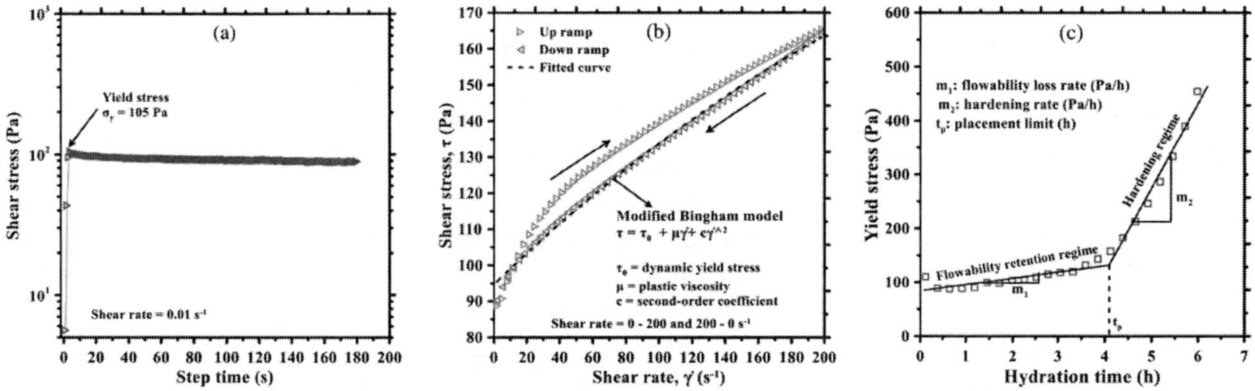

Figure 2. Typical representation of the CST pastes rheological process and quantifying metrics: (a) static yield stress obtained at 0.01 s^{-1}, (b) dynamic yield stress and plastic viscosity obtained by fitting the down ramp curve with a modified Bingham model, and (c) the evolution of static or dynamic yield stress as a function of hydration time, and the associated workability metrics.

Result and discussion

FTIR and XRD characterization of OPC, struvite powders, and CST samples

The FTIR spectra of the hydrated CST pastes and control are identical (Fig. 3), suggesting that no additional hydration products are formed with struvite addition to OPC; instead, the former aided hydrates formation by diluting cement particles in paste, providing relatively extra space for the nucleation of calcium silicate hydrate (C–S–H) gel [8,9]. The stretching and bending O–H bands at 3700-3300 and 1700-1600 cm^{-1}, respectively, due to portlandite or water, the stretching S–O band between 1200 and 1050 cm^{-1} due to gypsum or ettringite in the unhydrated cement or hydrated products, respectively, the stretching and bending C–O bands at 1486–1400 and 900–700 cm^{-1}, respectively, due to CO_3^{2-}, and the stretching and bending Si–O bands at 1000–900 and 800–400 cm^{-1}, respectively, due to C–S–H were characteristics of hydrated OPC [10–12]. Furthermore, the suppression of the prominent struvite absorption peaks (comparing struvite powder vs hydrated CST samples: N–H, O–H, and P–O) suggests that they must have been masked by the absorption peaks of the cement hydration products, bolstering the fact that they only serve as the surface for nucleation and growth of hydration products. XRD data (Fig. 4) further confirmed this hypothesis, showing that struvite remains in the CST samples irrespective of curing age. Therefore, the struvite absorption peaks did not disappear but were concealed by the cement hydration products.

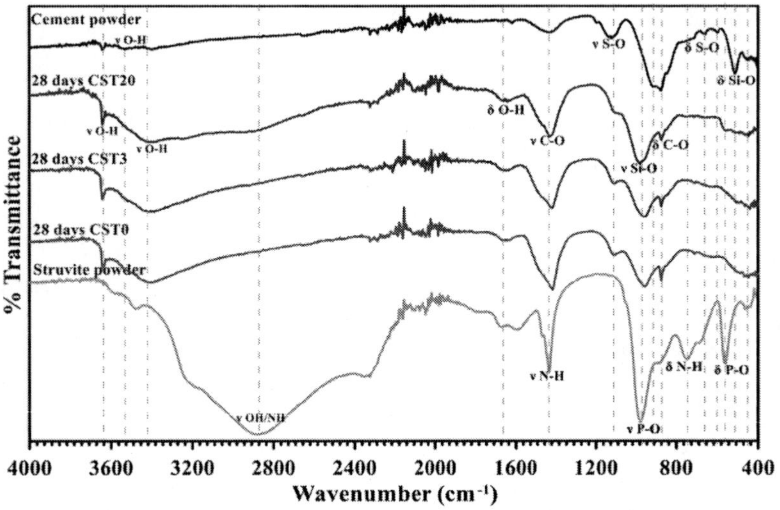

Figure 3. FTIR spectral bands of OPC and struvite powders and CST samples

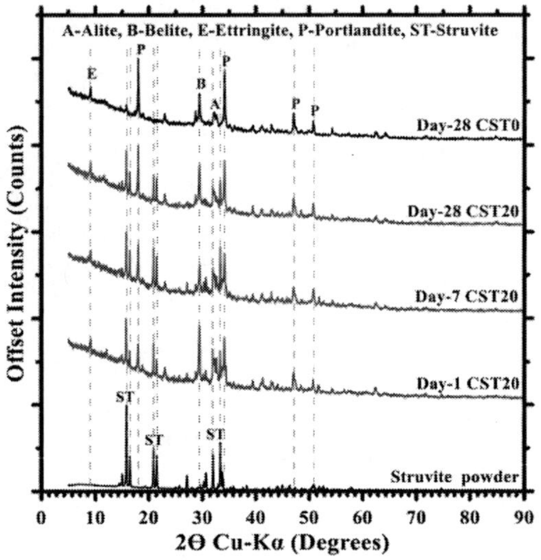

Figure 4. Phase assemblage of struvite and the CST hydration products

Micro X-ray computed tomography (XCT) and 3D image analysis for porosity

This study employed the XCT technique to study the porosity in CST pastes (CST0 and CST10) after 1 and 28 curing days. Fig. 5 shows the spatial distribution of the pores in the 3D CST cubes. Although the percentage of connected pores is more significant in CST10 than in the control, the volume occupied by the pores in the CST10 bulk cube is almost half that occupied in the control cube (Fig. 6a). Figures 6b, c, and d show the pore volume distribution, mean Feret diameter, and maximum Feret diameter, respectively, of CST0 and CST10 after 28 days. It is evident from the graphs that the addition of ST minimized the number of pores, the volume of pores, and Feret diameters.

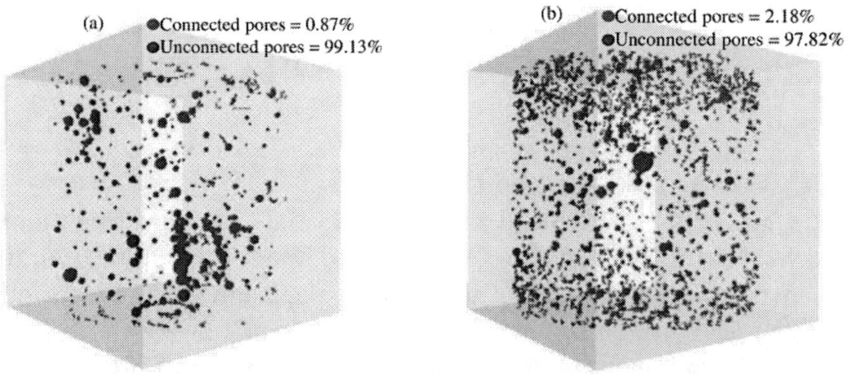

Figure 5. Porosity connectivity of (a) CST0 and (b) CST10 at 28 curing days.

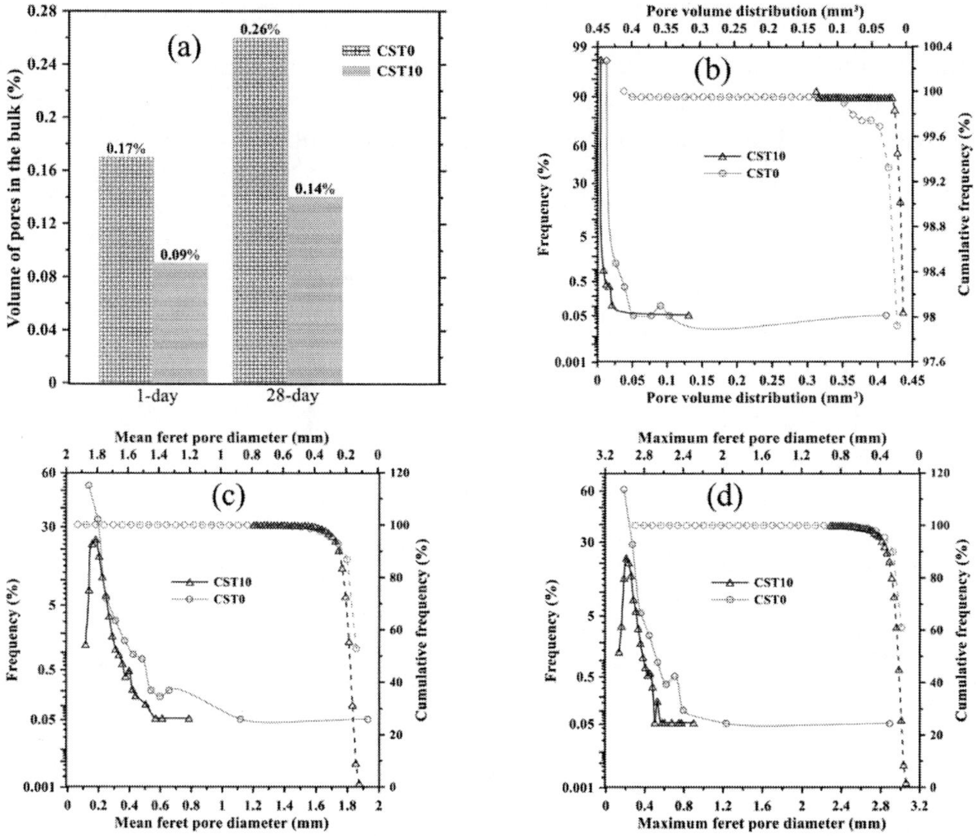

Figure 6. Volume of pores in the bulk (a) and porosity metrics (b: pore volume, c: mean Feret diameter, and d: maximum Feret diameter) of CST0 and CST10 at 28 curing days.

Mechanical properties of CST pastes

The FTIR and XRD results have provided the basis for considering the enhancement of cement hydration products by adding struvite. In addition, the porosity test indicates that the control paste, CST0, possesses more pores with higher sizes than the struvite-infused paste, CST10. Porosity reduction leads to an increase in the mechanical performance and durability of cementitious materials [13]. These tests corroborate the high mechanical performance of CST pastes. Figure 7 shows an increase in compressive strength of CST pastes with a percentage increase in ST. The impressive performance of CST pastes is evident with the achievement of >100% strength activity index up to 20% replacement (Fig. 7b). The flexural strength also increased with an increasing struvite addition (Fig. 7c).

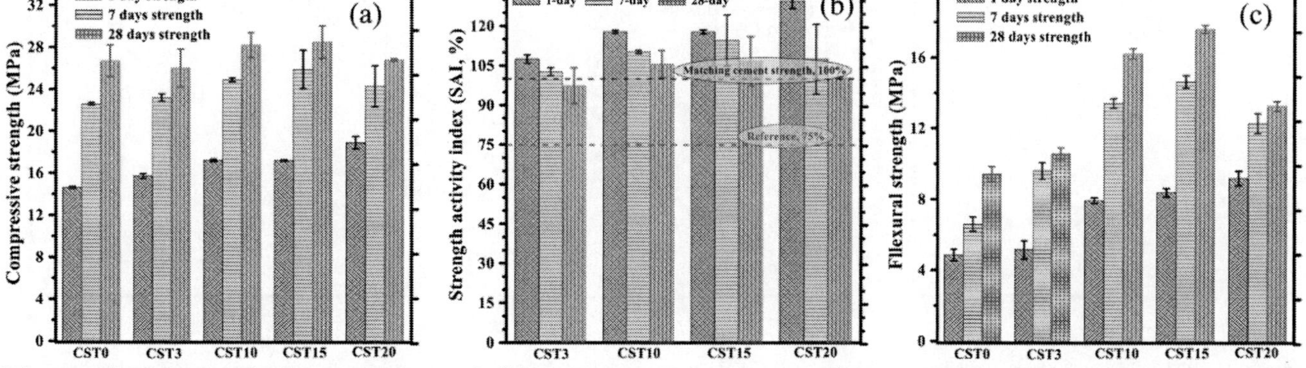

Figure 7. Mechanical properties of CST pastes: (a) compressive strength, (b) compressive strength activity index, and (c) flexural strength.

Hydration kinetics of CST pastes

Figure 8 shows an initial increase in heat evolution with increasing struvite addition in the first 2 hours of hydration. After that, a prolonged induction period before the acceleration phase culminates in a decrease in the heat evolution of the CST pastes. The heat reduction corresponds to the increase in percentage ST addition, implying that the CST pastes may have reduced risk of cracking, improved durability, and higher resistance to chemical shrinkage.

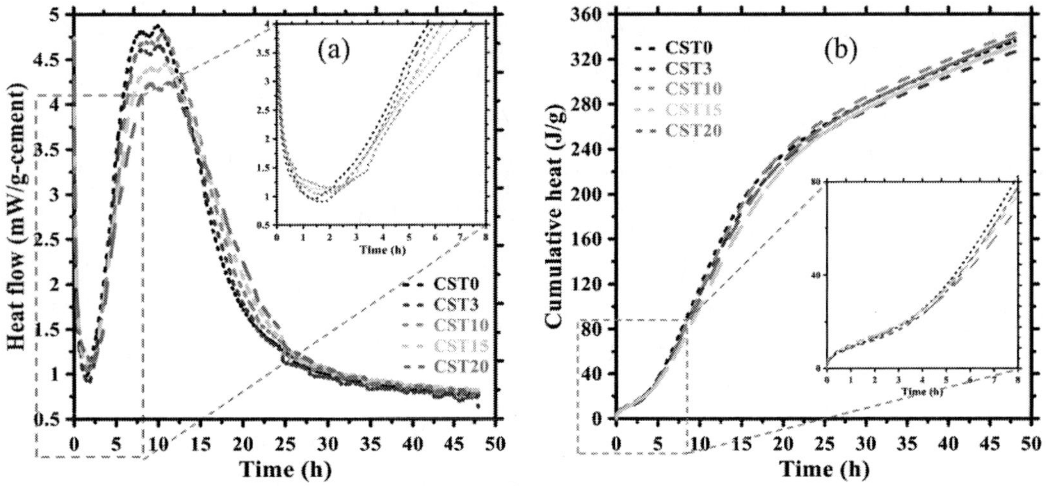

Figure 8. Hydration kinetics of CST pastes showing (a) heat evolution and (b) cumulative heat evolution

Rheological performance of CST pastes

Important rheological properties of CST paste, which provide critical information about its behavior in the fresh state, are elucidated in Fig. 9. These are also essential for various aspects of concrete production and application, such as mixing, pumping, placing, and finishing. Figures 9a and b depict increased static and dynamic yield stresses with increasing ST addition. Static yield stress is a critical rheological parameter, with sufficiently high static yield stress being a precondition for 3D concrete [14]. It is observed that the dynamic yield stress of CST pastes increased along with the static yield stress, indicating a structural build-up of hydration products with a commensurate increase in strength, as shown in Fig. 7. However, there is a decrease in plastic viscosity with an increasing percentage of ST addition (Fig. 9c). This is a rare but significantly beneficial property of CST pastes, which do not require flow modifiers to achieve optimal mixing, transportation, and casting fluidity.

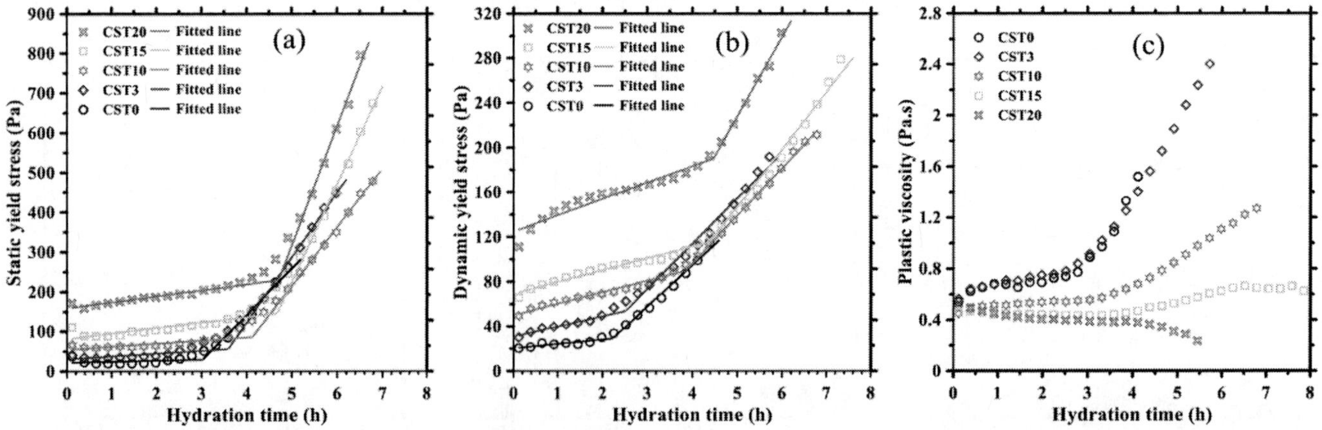

Figure 9. Rheological characterization of CST pastes displaying (a) static yield stress evolution, (b) dynamic yield stress, and (c) plastic viscosity

Furthermore, modeling the static yield stress and dynamic yield stress curves of the CST pastes resulted in the workability metrics presented in Fig. 10. The placement period of the CST pastes increased with increasing substitution with ST. This trend is bolstered by the delayed hydration observed in Fig. 8. That means that by increasing the substitution of OPC with ST, the hydration rate of CST pastes was reduced. This also supported the higher strength of the CST pastes because uniform heterogeneous nucleation and growth often lead to dense and compacted hydrated products. Although the CST pastes maintained an increasing flowability loss rate, which was expected due to high yield stress, the hardening rate obtained from the dynamic yield stress curve was lower. This is advantageous for the mixing and transportation of concrete.

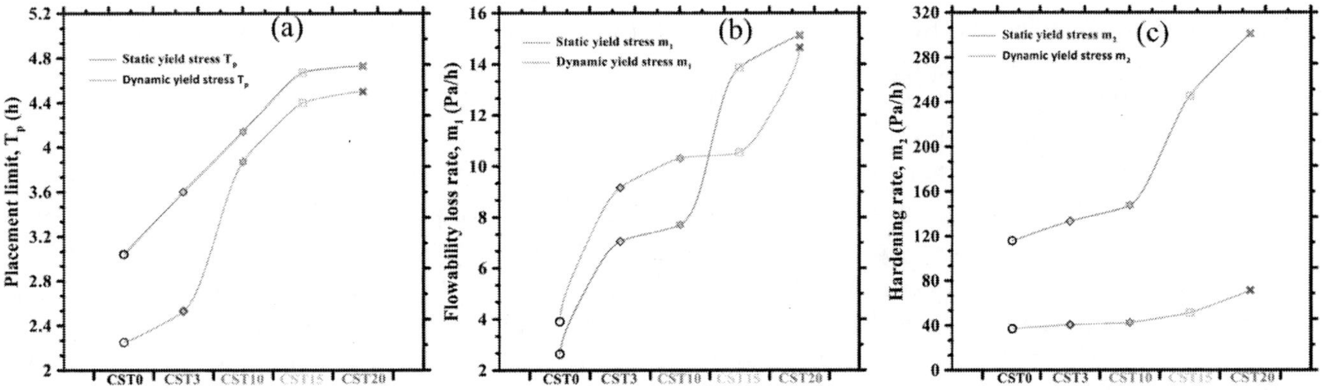

Figure 10. Workability metrics of CST paste indicating (a) placement limit, (b) flowability loss rate, and (c) hardening rate.

Conclusion

Struvite has been shown as a potential candidate for OPC substitution to reduce carbon emissions associated with OPC production. The CST paste achieved a strength activity index greater than 100% for up to 20% substitution, implying that they not only met the requirements for cement substitutes but also surpassed the strength of OPC. The results of the FTIR and XRF analyses point to the fact that struvite provides enabling surfaces for nucleation and the growth of cement hydration products and remains stable. Having reduced porosity suggests that concretes made with CST blend will possess high resistance to chemical attacks. In addition, the reduced hydration heat offers additional resistance to chemical shrinkage and thermal cracking. More importantly, reducing cement paste plastic viscosity through ST addition– without adding flow modifiers– is a promising property of CST blend, which will reduce construction costs while prompting a cleaner environment.

References

[1] Supriya, R. Chaudhury, U. Sharma, P.C. Thapliyal, L.P. Singh, Low-CO2 emission strategies to achieve net zero target in cement sector, Journal of Cleaner Production 417 (2023) 137466. https://doi.org/10.1016/j.jclepro.2023.137466.

[2] P. Jittamaro, B. Maho, P. Pongsopha, D. Nicomrat, S. Jamnam, N. Makul, G. Sua-iam, Enhancing the usability of electronic waste fibers in high-performance self-compacting mortar incorporating corn cob ash and silica fume: Fresh and hardened properties, Construction and Building Materials 416 (2024) 135194. https://doi.org/10.1016/j.conbuildmat.2024.135194.

[3] V.B. Aguilar-Pozo, J.M. Chimenos, S. Peña-Picola, K. Olaciregui-Arizmendi, B. Elduayen-Echave, E. Ayesa, S. Astals, Process performance and precipitate quality of phosphorus recovery by struvite precipitation in a fluidized bed reactor using a MgO industrial by-product, Journal of Environmental Chemical Engineering 12 (2024) 112764. https://doi.org/10.1016/j.jece.2024.112764.

[4] W. Mao, C. Cao, X. Li, J. Qian, C. You, Preparation of Magnesium Ammonium Phosphate Mortar by Manufactured Limestone Sand Using Compound Defoaming Agents for Improved Strength and Impermeability, Buildings 12 (2022) 267. https://doi.org/10.3390/buildings12030267.
[5] Y. Liu, B. Chen, B. Dong, Y. Wang, F. Xing, Influence mechanisms of fly ash in magnesium ammonium phosphate cement, Construction and Building Materials 314 (2022) 125581. https://doi.org/10.1016/j.conbuildmat.2021.125581.
[6] V.B. Aguilar-Pozo, J.M. Chimenos, B. Elduayen-Echave, K. Olaciregui-Arizmendi, A. López, J. Gómez, M. Guembe, I. García, E. Ayesa, S. Astals, Struvite precipitation in wastewater treatment plants anaerobic digestion supernatants using a magnesium oxide by-product, Science of The Total Environment 890 (2023) 164084. https://doi.org/10.1016/j.scitotenv.2023.164084.
[7] J. Hövelmann, C.V. Putnis, In Situ Nanoscale Imaging of Struvite Formation during the Dissolution of Natural Brucite: Implications for Phosphorus Recovery from Wastewaters, Environ. Sci. Technol. 50 (2016) 13032–13041. https://doi.org/10.1021/acs.est.6b04623.
[8] K.L. Scrivener, A. Nonat, Hydration of cementitious materials, present and future, Cement and Concrete Research 41 (2011) 651–665. https://doi.org/10.1016/j.cemconres.2011.03.026.
[9] M. Wu, S. Sui, Y. Zhang, Y. Jia, W. She, Z. Liu, Y. Yang, Analyzing the filler and activity effect of fly ash and slag on the early hydration of blended cement based on calorimetric test, Construction and Building Materials 276 (2021) 122201. https://doi.org/10.1016/j.conbuildmat.2020.122201.
[10] R.C. Mendoza, J.O. Grande, M.N. Acda, Effect of Keratin Fibers on Setting and Hydration Characteristics of Portland Cement, Journal of Natural Fibers 18 (2021) 1801–1808. https://doi.org/10.1080/15440478.2019.1701604.
[11] S. Suman Lai, Characterization of Some Cement Samples of Nepal Using FTIR Spectroscopy, International Journal of Advanced Research in Chemical Science 5 (2018). https://doi.org/10.20431/2349-0403.0507004.
[12] M. Derrick, D. Stulik, J. Landry, Infrared Spectroscopy in Conservation Science, Scientific Tools for Conservation. Los Angeles, CA: Getty Conservation Institute, 1999. http://hdl.handle.net/10020/gci_pubs/infrared_spectroscopy.
[13] J.S. Vasconcellos, Y.S.B. Fraga, J.H.D.S. Rêgo, P.P.C. Sartoratto, M.F. Rojas, Hydration, mechanical performance and porosity of Portland cement pastes with functionalized nanosilica with APTES, Developments in the Built Environment 14 (2023) 100157. https://doi.org/10.1016/j.dibe.2023.100157.
[14] C. Zhang, Z. Deng, C. Chen, Y. Zhang, V. Mechtcherine, Z. Sun, Predicting the static yield stress of 3D printable concrete based on flowability of paste and thickness of excess paste layer, Cement and Concrete Composites 129 (2022) 104494. https://doi.org/10.1016/j.cemconcomp.2022.104494.

682e Anaerobic Treatment and Valorization of Aqueous Products from Hydrothermal Liquefaction of Wet Organic Wastes

Yalin Li, Xavier Lennard, Jason Chiang, Md. Nizam Uddin, L. Stetson Rowles, Hyejeong Kwon, Daqian Jiang

Proper management of wet organic wastes (e.g., food waste) is receiving growing attention due to their increasing quantity. However, most of this waste stream is currently sent to the landfill, contributing to a significant amount of greenhouse gas (GHG) emissions. Four states (Connecticut, Massachusetts, New Jersey, and Vermont) in the US have enacted policies limiting food waste to landfills; thus, additional handling and disposal routes are of urgent interest. Hydrothermal liquefaction (HTL) is an emerging technology that leverages the unique properties of water at elevated temperatures and pressures to convert wet organic wastes into crude fuel and chemical products. While suggested to be a cost-effective waste management approach with much lower GHG emissions for certain feedstocks (e.g., sewage sludge compared to landfill, incineration, and land application; **Figure A**), there are still needs to further improve the economic performance to address the financial risks associated with commercial deployment, especially considering the capital-intensive nature of such systems.

One strategy to enhance cost competitiveness is to improve the valorization of HTL coproducts. Unlike the main energy-dense biocrude product, HTL aqueous product (HTL-AP) has received much less attention, largely due to its complexity that prevents low-cost valorization yet necessitates proper treatment before discharge or recycling. At the same time, the high organic contents (in terms of chemical oxygen demand, COD) of HTL-AP makes anaerobic digestion (AD) a potential route for simultaneous energy production and treatment. This presentation will discuss findings from the anaerobic treatment of HTL-AP of different feedstocks (ground coffee and food waste from the university dining hall) and reaction conditions (temperature and homogeneous additives), where HTL experiments have shown great variation in yields of biocrude and hydrochar (residuals; **Figure B**). Two mesophilic anaerobic systems were explored, including one batch AD system and one sequencing batch bio-electrochemical AD system (BEAD) where electricity was used to overcome the inhibitory effects from the complex aqueous products (e.g., ammonia, aldehydes, N-heterocyclic compounds, aromatic compounds). The batch experiments were performed in 250 mL reactors with HTL-AP diluted to a COD of 1 $g \cdot L^{-1}$. For the sequencing experiments, three parallel tests were set up in 1 L reactors: one control test with acetate, one AD test with acetate-HTL-AP mixture but 0 voltage, and one BEAD test with acetate-HTL-AP mixture and 0.5 V of voltage. Methane yields from both batch and sequencing systems were monitored daily. Finally, cost and environmental impact implications of the experimental findings were interpreted through processed-based techno-economic analysis (TEA) and life cycle assessment (LCA) on QSDsan, an open-source platform for quantitative sustainable design of sanitation and resource recovery technologies. Two alternatives of an integrated hydrothermal waste valorization system consist of HTL, biocrude upgrading, and co-product valorization were considered: one employing chemical valorization of HTL-AP through catalytic hydrothermal gasification, while the other employing biological valorization through AD (**Figure C**).

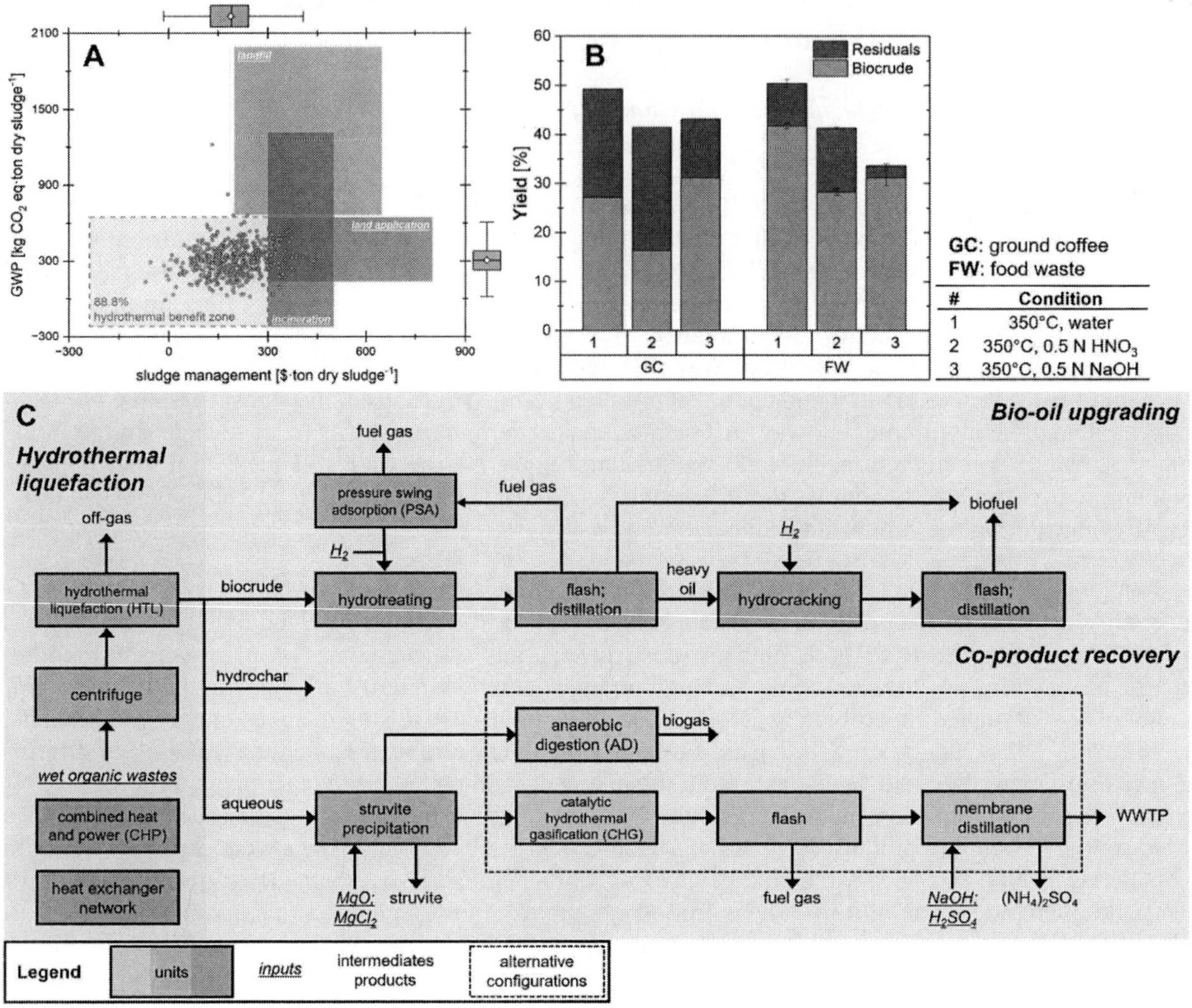
A
GWP [kg CO2 eq·ton dry sludge-1]
sludge management [$·ton dry sludge-1]
landfill
land application
incineration
88.8%
hydrothermal benefit zone
B
Yield [%]
Residuals
Biocrude
GC
FW
GC: ground coffee
FW: food waste
Condition
1 350°C, water
2 350°C, 0.5 N HNO3
3 350°C, 0.5 N NaOH
C
Hydrothermal liquefaction
Bio-oil upgrading
Co-product recovery
off-gas
hydrothermal liquefaction (HTL)
centrifuge
wet organic wastes
combined heat and power (CHP)
heat exchanger network
biocrude
hydrochar
aqueous
fuel gas
pressure swing adsorption (PSA)
H2
hydrotreating
flash; distillation
heavy oil
hydrocracking
biofuel
anaerobic digestion (AD)
biogas
struvite precipitation
MgO; MgCl2
struvite
catalytic hydrothermal gasification (CHG)
flash
membrane distillation
NaOH; H2SO4
(NH4)2SO4
WWTP
Legend
units
inputs
intermediates products
alternative configurations

682f Improving Food Waste Anaerobic Digestion Efficiency with Biochar in Decentralized System

Yen Wah Tong

Long-term anaerobic digestion of food waste encounters challenges like feedstock overloading, leading to volatile fatty acids accumulation and process disruption. Although biochar-enhanced anaerobic digestion shows promise in addressing these issues, its effectiveness in small-scale labs limits its adoption in decentralized anaerobic digestion systems. This study investigates the efficacy of biochar supplementation in such a decentralized setting, revealing that a modest biochar quantity (0.055 g/L) improves anaerobic digestion under ambient conditions (29°C). Within 36 days, volatile fatty acids inhibition decreases to a safe level of 1195 mg/L. The normalized specific biogas yield increases by 50% compared to the previous control study, reaching 1.40 ± 0.74 m^3/kgVS. After achieving stability, anaerobic digestion at mesophilic conditions (35°C) without biochar sees a further 27.8% yield increase to 1.79 ± 0.83 m^3/kgVS. Additionally, methane biogas concentration rises to 60-62%, compared to 56.9 ± 2.9% in the control study. In terms of microbiome composition, biochar enhances *Methanosaeta* genus methanogens by 30%, facilitating direct interspecies electron transfer and enhancing archaea-bacteria symbiosis. These findings highlight biochar's potential in decentralized biogas facilities and sustainable waste management, offering a replicable model for closing the food waste loop and advancing the bioeconomy.

128a Experimental Investigations and Mechanistic Insights into the Sorptive Removal of Elemental Gas Phase Mercury (Hg^0) from Syngas over Pt-Based Sorbents

Dwijraj Mhatre, Divesh Bhatia

Introduction

Mercury, a toxic heavy metal, is a trace component in carbonaceous feedstocks, viz., coal, biomass, and municipal solid wastes, and gets released during their combustion and gasification in its elemental form (Hg^0). The utilization of syngas derived from the gasification of carbonaceous feedstocks to produce valuable chemicals, demands the syngas to be free from Hg^0 which can poison the sensitive catalyst (s) employed downstream. Hg^0 is volatile and insoluble in water and cannot be trapped using conventional air pollution control equipment, due to which an alternative technology is required for its removal. Additionally, the capture of Hg^0 from syngas is favored at elevated temperatures (>200 °C) to improve the overall thermal efficiency of the interconnected gasification and chemical processes. However, the commercially available carbon and sulfur-based adsorbents are inefficient for the removal of Hg^0 from syngas at high temperatures. DFT coupled ΔG calculations have shown Hg^0 adsorption to be feasible on Pt surfaces at high temperatures (>200 °C) [1].

In the present work, a series of Pt-based sorbents are synthesized using different support materials, and their efficacy in Hg^0 removal at high temperatures is evaluated. XPS analysis of the fresh and spent sorbents is used to investigate the mechanism for Hg^0 removal, and two different support-dependent pathways are proposed using results from XPS, XRD, and H_2-TPR analysis. Experiments are performed to elucidate the effect of temperature and the role of different syngas components on Hg^0 removal. Further, the effect of syngas components H_2 and H_2O is shown to vary over different sorbents, and the negative effect of CO on Hg^0 removal over all sorbents is explained using in-situ DRIFTS studies.

Methodology

The Pt-based monometallic adsorbents shortlisted from DFT calculations are prepared using the incipient wetness method using doped aluminas (SIRAL-40, SASOL), metal oxides (Ceria, zirconia, titania) and silica-based molecular sieve (SBA-15) as support materials. The performance of these sorbents is investigated using a bench-scale reactor system at high temperatures (200 to 350 °C) using a Hg^0 feed concentration of 25 ± 6 $\mu g/m^3$ and by maintaining a constant gas hourly space velocity (GHSV) of 1.2×10^5 h^{-1}. The physio-chemical properties of the sorbent are determined using BET), XRD, and H_2-TPR. XPS analysis of the spent sorbents exposed to Hg^0 for 10 hours is performed to investigate the Hg^0 removal mechanism. The effect of various syngas components viz., CO, CO_2, H_2, and H_2O on Hg^0 removal performance at 270 °C is shown, and the effect of CO on Hg^0 removal performance is investigated using in-situ DRIFTS.

Results and Discussion

The narrow spectrum of Pt 4f for fresh and spent Pt/SBA-15 is given in Fig. 1 (a), wherein the two peaks at 71.8 eV and 75.1 eV indicate metallic Pt (Pt^0). Following Hg^0 adsorption, the Pt 4f spectra of spent Pt/SBA-15 exhibit a gradual shift towards lower binding energy, suggesting alterations in the Pt states. Furthermore, deconvolution of the XPS spectra reveals two new peaks at 71.1 eV and 74.4 eV on spent Pt/SBA-15, indicative of the formation of a Pt-Hg amalgam after Hg^0 adsorption [2]. Similarly, the XPS analysis of the other Pt-based sorbents is performed and Hg^0 removal over all sorbents is proposed to occur by Hg^0 adsorption and formation of a Hg-Pt amalgam. From the XPS spectra of O 1s on fresh and spent Pt/ZrO_2, a decrease in the proportion of adsorbed oxygen (O_β) is observed after Hg^0 adsorption (Fig. 1 (b)), based on which Hg^0 oxidation is proposed to occur. Similar inferences are obtained for Pt/CeO_2 as well. Characterization results from XRD and H_2-TPR indicate the presence of Pt nanoparticles in the elemental state on Pt/SBA-15, Pt/S-40, and Pt/TiO_2, and the presence of Pt in the oxidized form on Pt/ZrO_2 and Pt/CeO_2. The correlation of these characterization techniques with the adsorption activity would be described in the full manuscript.

The initial Hg^0 removal efficiency of various Pt-based sorbents is plotted in Fig. 2 (a). Pt/SBA-15, Pt/SIRAL-40 (S-40), and Pt/ZrO_2 demonstrate Hg^0 removal efficiencies above 90% for temperatures up to 250 °C. With an increase in the adsorption temperature to 270 °C, the removal efficiency of Pt/S-40 and Pt/ZrO_2 declines to 91.0 % and 85.4 % respectively, whereas the efficiency of Pt/SBA-15 remains above 97.6 %. However, at a higher adsorption temperature of 350 °C, the removal efficiency of Pt/ZrO_2 and Pt/CeO_2 is higher than Pt/SBA-15 which could be explained by the occurrence of Hg^0 oxidation. Pt/SBA-15 is evaluated over varying temperatures for longer time durations, and the results are plotted in Fig. 2 (b). At 200 °C, the efficiency of Hg^0 removal declines below 95% after 55 h, whereas this drop occurs much earlier (within 0.68 h) at a higher adsorption temperature of 350 °C. The decline in efficiency is attributed to the reduction in Hg^0 solubility in Pt and the initiation of de-amalgamation. The Hg^0 removal efficiency over Pt/SBA-15 at 270 °C in N_2 is comparable to that reported for Pd-based sorbents [3], making it suitable for Hg^0 removal from syngas at high temperatures. Further, a one-dimensional two-phase transient model is developed to predict the spatio-temporal features for Hg^0 sorption over Pt/SBA-15 under isothermal conditions. The model considers the adsorption of Hg^0 on the surface of Pt followed by amalgam formation. The model predicts Hg^0 sorption over a range of temperatures and also the decrease in the adsorption capacity with an increase in temperature from 200 to 350 °C.

The Hg^0 removal efficiency at 270 °C for various compositions of syngas is shown in Fig. 3. The adsorption efficiency over Pt/S-40, Pt/ZrO_2, Pt/TiO_2, and Pt/CeO_2 reduces significantly in the presence of syngas components CO, H_2 and H_2O thus showing their inhibitive effect towards Hg^0 adsorption. In contrast, the adsorption efficiency of Pt/SBA-15 is retained above 90% in the presence of H_2 and H_2O. Thus, mesoporous and hydrophobic silica-based SBA-15 is highlighted as a promising support material for Pt-based sorbents in eliminating Hg^0 from syngas at high temperatures in the presence of H_2O. In-situ DRIFTS experiments with CO in the feed are performed, which show the adsorption of CO and explain its competitive adsorption with Hg^0 (Fig. 4 (a)). On stopping CO and purging N_2, the CO bands completely disappear indicating the complete desorption of CO. This makes the Pt^0 sites available for Hg^0 adsorption thereby explaining the restoration of Hg^0 removal efficiency over Pt/SBA-15 after stopping the flow of CO (Fig. 4 (b)).

Significance of the work

Pt-based adsorbents for the removal of Hg^0 at high temperatures exceeding 200 °C are developed in contrast to commercial adsorbents which work at much lower temperatures. A combination of reactor experiments and characterization (in-situ DRIFTS, XPS, XRD, and H_2-TPR) is used to propose a support and temperature-dependent Hg^0 removal mechanism.

References

[1] D. Mhatre, D. Bhatia, Insights into the adsorption, alloy formation, and poisoning effects of Hg on monometallic and bimetallic adsorbents, Langmuir. 38 (2022) 6841–6859.

[2] G.W. Wu, S. Bin He, H.P. Peng, H.H. Deng, A.L. Liu, X.H. Lin, X.H. Xia, W. Chen, Citrate-capped platinum nanoparticle as a smart probe for ultrasensitive mercury sensing, Anal. Chem. 86 (2014) 10955–10960.

[3] L. Han, Q. Li, S. Chen, W. Xie, W. Bao, L. Chang, J. Wang, A magnetically recoverable Fe_3O_4-NH_2-Pd sorbent for capture of mercury from coal derived fuel gas, Sci. Rep. 7 (2017) 2–11.

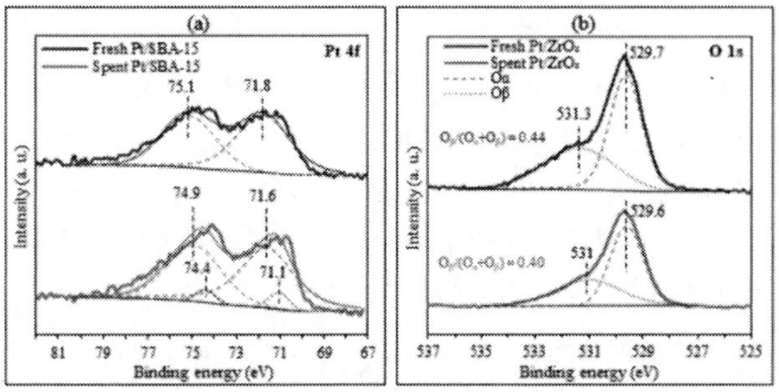

Fig. 1. (a) XPS spectra of Pt 4f on fresh and spent Pt/SBA-15, and (b) XPS spectra of O 1s on fresh and spent Pt/ZrO_2.

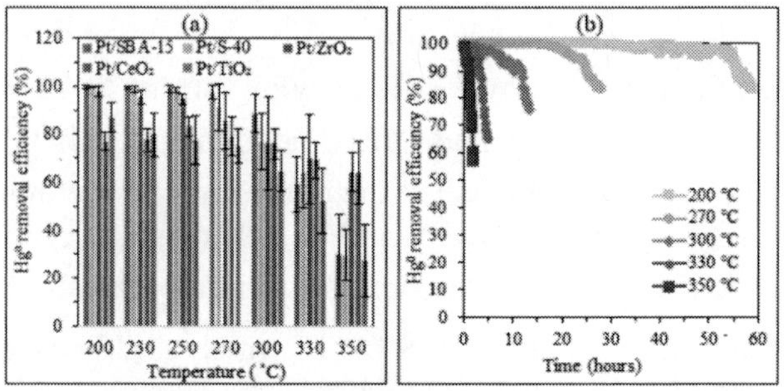

Fig. 2. (a) Hg^0 removal efficiency of Pt-based sorbents at different temperatures (b) Effect of temperature on Hg^0 removal efficiency of Pt/SBA-15.

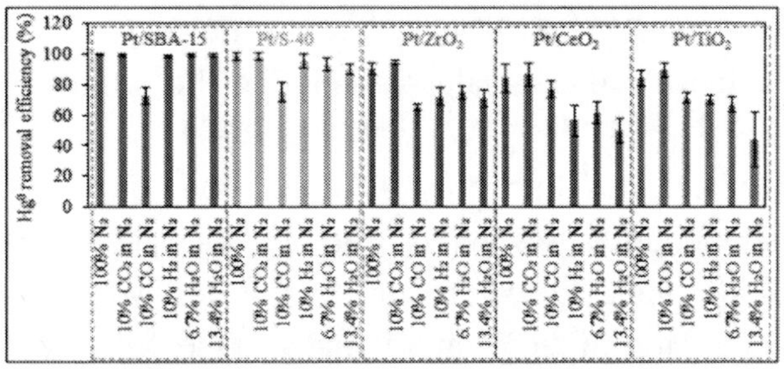

Fig. 3. Effect of syngas components on Hg^0 removal efficiency at 270 °C.

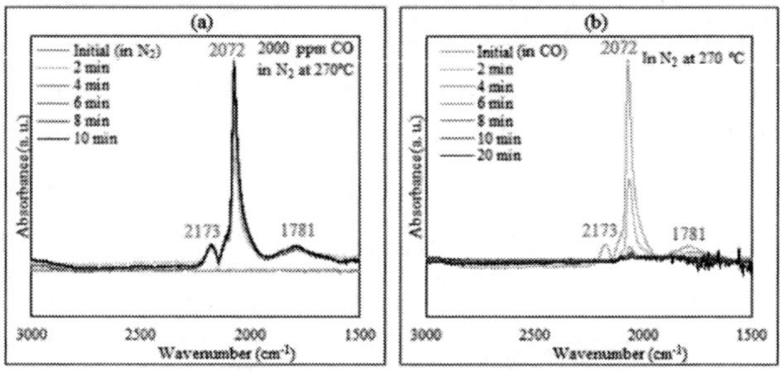

Fig. 4. In-situ DRIFTS spectra at 270 °C over Pt/SBA-15 during (a) CO exposure, and (b) N_2 purging after CO exposure.

128b Kinetics of Flocculation of Cyanobacteria Using Alum and Moringa Biocoagulant

Temitope Orimolade, Ngoc-Tram Le, Lyle Trimble, Bandaru V. Ramarao, Sitaraman Krishnan

There is great interest in purifying water bodies contaminated by *Microcystis aeruginosa*, a cyanobacteria responsible for harmful algal blooms and the release of microcystin toxins. Flocculation is the preferred approach for separating these cells, and flocculants that are more environmentally benign than conventional metal salt coagulants are desired. The seeds and extracts of *Moringa oleifera* are known to be quite effective flocculants in purifying turbid water. However, a detailed investigation of the effect of flocculant dosing on the flocculation kinetics and the floc size, as well as the comparison of the flocculation properties of *Moringa* with alum, is lacking in the literature.

In the work reported herein, laser diffraction analysis was used to investigate the kinetics of flocculation of *M. aeruginosa* using *M. oleifera* in three forms: powdered seeds, aqueous extracts of the powdered seeds containing the active protein coagulant, and aqueous extracts of the deoiled seeds. The flocculation process was carried out at various concentrations of the cyanobacterial cells and the coagulant. The results are compared with those from similar experiments using alum as the coagulant. The laser diffraction data were analyzed for the time evolution of the floc's size and fractal dimension, allowing the flocculation process to be investigated in much greater detail than possible using conventional turbidity measurements.

128c Effects of Microbial Soil Amendments to Improve Soil Water Availability and Denitrification.

Moises Gutierrez, Ryan Hansen

This work investigates the use of *Bacillus subtilis* as a soil additive that can improve water retention in agricultural soils, reducing water demands during periods of drought stress. In the soil, *B. subtilis* can produce cyclic lipopeptides as secondary metabolites that act as surfactants, breaking down soil water repellency and generating hydrophilic soils, enhancing soil water-holding capacity. In the prior work, *B. subtilis* was added to sandy loam soils in a lab-scale system; here, it was found that soil inoculated with *B. subtilis* can decrease the soil moisture threshold, increasing soil moisture availability. In the current work, we investigate the effect that different forms of inorganic nitrogen (N) have on *B. subtlis* mediated soil wetting. Evaporation experiments were conducted by treating *B. subtilis* inoculated soils with various forms of nitrogen (NH_4NO_3, NH_4Cl, $NaNO_3$), representative of different nitrogen-based fertilizers. Evaporation rates and soil water retention levels were then measured while heating the soils. Results indicated that soils treated with NH_4NO_3 retained significantly more water than untreated soils and soils treated with other forms of nitrogen. Inorganic nitrogen levels (ammonium, nitrate) were quantified during each experiment using KCl extraction to show that soils treated with NH_4NO_3 undergo the most intense denitrification process, reducing nitrate levels well below initial levels, despite adding more nitrates during the inoculation. These findings indicate that soils amended with NH_4NO_3 provided the highest gains in soil water retention and that inoculating soils with *B. subtilis* could deliver multiple benefits to stressed agriculture soils by both increasing soil water holding capacities and by providing denitrification to reduce nitrogen runoff and loss.

128d Fungal Contributions to the Birch Effect: Insights from Microfluidics in Dryland Ecosystems

Yi-Syuan Guo, Karl Weitz, Arunima Bhattacharjee, Aramy Truong, Adam Ryan, Leslie M. Shor, Mary S Lipton

Dryland ecosystems represent a significant biome on Earth, but they are highly susceptible to the impacts of drought and rewetting events that can reduce species diversity and agricultural productivity. With these ecosystems expanding due to climate change, understanding the effects of such events on the soil microbiome is crucial for accurately predicting and mitigating the impact of climate change on ecosystems and carbon cycling. The Birch Effect, arising from drying and rewetting cycles, leads to the efflux of carbon dioxide (CO_2) which can be monitored via real-time mass spectrometry (RTMS). While soil fungi play a crucial role in water retention during extended droughts, their specific contribution to CO_2 efflux during dry-rewetting cycles and the underlying pore-scale biogeochemistry remains unclear. In this study, we propose an investigation of the Birch Effect using soil micromodels to simulate drought and rewetting cycles. Initially, these micromodels were inoculated with fungi to observe their growth dynamics with and without access to minerals. Subsequently, the micromodels underwent drying to evaluate the impact of dried fungi on moisture retention within soil pores. Following a sufficient drying period, RTMS was employed to quantify CO_2 release upon rewetting the micromodels with deionized water. This approach offers insights into the fungal hydraulic lift and the consequent fungal contributions to the Birch Effect, elucidating the critical role of soil fungi in ecosystems.

128e System Dynamics Modeling for Sustainable Urban Development in the City of Cape Town: A Water-Energy-Food Nexus Approach

Jafaru Egieya, Yumna Parker, Viola Hofmann, Bassel Daher, Efstratios Pistikopoulos, Johann Gorgens, Neill Goosen

Between 2015 and 2018, the City of Cape Town (CoCT), South Africa, confronted an unprecedented water scarcity crisis known as Day Zero, marked by the near-complete depletion of its municipal water supply [1]. This critical situation was precipitated by a significant decline in rainfall, leading to one of the severest droughts on record, and was further exacerbated by rapid population growth and urbanization. These factors, coupled with increased urban energy consumption and food demand, posed a significant threat to the city's water resources.

In response to this complex crisis, this study adopted a system dynamics modeling (SDM) approach [2] to explore the intricate interdependencies among water, energy, and food (WEF nexus). Utilizing Stella Architect, a comprehensive model was constructed as a stock and flow diagram, encompassing five interconnected subsystems to assess the dynamics of water utilization, production, and the broader implications for energy supply and food production.

The study aims to explore the long-term impacts of interconnected variables on Cape Town's water security, with a particular emphasis on understanding the influence of environmental changes and urban development patterns. Four distinct scenarios—business-as-usual (BAU), climate change integration with energy transition (CCI), water integration schemes (WIS), and food habit change (FHC)—are examined over a 20-year modeling period. Notably, the water integration scheme emerges as the most promising, offering a significant increase in water availability compared to other scenarios, with a minimum of 20% improvement demonstrated.

These insights are poised to offer valuable guidance for policymakers and urban planners in CoCT and beyond, grappling with similar water-related challenges. By leveraging these findings, stakeholders can better anticipate and address potential crises, fostering sustainable urban water management strategies that are crucial for the resilience and prosperity of cities worldwide.

References

[1] Hachaichi M., Egieya J.M. 2023. Water-food-energy nexus in global cities: addressing complex urban interdependencies. *Water Resources Management.* 4, 1911-1825,

[2] Wen, C., Dong, W., Zhang, Q., He, N., Li, T. 2022. A system dynamics model to simulate the water-energy-food nexus of resource-based regions: A case study in Daqing City, China. Science of the Total Environment. 806, 150497.

128f Robust Insights into CO2 Hydrate Kinetics in Brine-Saturated Sediments Using Experimental and Machine Learning Approaches for Ccus Application

Vikas Dhamu, M Fahed Qureshi, Praveen Linga

One of the potential Carbon Capture and Sequestration (CCS) methods is capturing CO_2 and injecting it into the oceanic sediments where CO_2 is converted into CO_2 gas hydrates. However, the presence of brine [NaCl] in sediments likely affects the kinetics of CO_2 hydrate formation significantly. To tackle this challenge, a real-world scenario was mimicked in this work, by creating artificial brine-rich different-sized sediment beds in a laboratory-scale reactor. Afterward, CO_2 was injected multiple times via injection tube directly into the sediment, and the CO_2 hydrate formation, dissociation, and morphological changes were recorded.

The experimental results show that the water-to-hydrate conversion [%] was estimated to be about highest in smaller size sediments [~60%] > both (smaller + larger sediments) dual [~35 %] > larger sediments [~20 %]. The visual observations show the smaller-size sediments show the haphazard isotropic advancement of whitish CO_2 hydrate whereas in the presence of large-sized sediments, various hydrate capping which trapped brine or CO_2 was observed. Moreover, using ~94,000 experimental data (% water to hydrate conversion, pressure, and temperature) a newly proposed CO_2 hydrate formation kinetic model was trained using a supervised ML-based algorithm to predict best-optimized parameters having the capability of forecasting CO_2 hydrate formation kinetics with %AARD of < 9%. This study's experimental and modelling results offer valuable insights that can play a pivotal role in advancing the CCS via hydrate technologies.

Acknowledgments

PL acknowledges the funding support from the Agency of Science, Technology and Research (A*STAR) under the low carbon energy research (LCER) funding initiatives (project ID: U2102d2010).

128g An L-Tryptophan and 1,3 Dioxane Synergic Study on CO_2 Hydrate Kinetic with Seawater: Applicable to CO_2 Sequestration Via Gas Hydrate

Vikas Dhamu, M Fahed Qureshi, Praveen Linga

One of the potential carbon capture and storage [CCS] techniques is capturing CO_2 emissions from the industry and injecting them into the oceanic sediments, where, under favorable conditions, they can be stored as gas hydrates. However, the high salinity of oceanic sediments hinders the CO_2 hydrate formation kinetics, which is a major challenge in hydrate-based CCS. To address this challenge, the fundamental study of CO_2 hydrate formation, dissociation, and morphology in seawater was carried out in the static and non-static mediums. Moreover, the effect of a kinetics hydrate promoter (L-tryptophan), thermodynamic hydrate promoter (1,3 dioxane), and synergic effects of kinetics+ thermodynamics hydrate promoters on CO2 hydrate kinetics was also analyzed. Furthermore, using Raman spectroscopy, CO_2 dissolution behavior was evaluated in seawater with 1,3 dioxane.

The experimental results show that the CO_2 uptake was higher for stirred medium [39 (±4) to 67 (±2) mmol/mol] compared to unstirred medium [21 (±6) to 30 (±5) mmol/mol] when different combinations of the promoters (L-tryptophan and 1,3 dioxane) were used. The morphological results indicate that CO_2 hydrate formed using 1,3 dioxane has different needle-like CO_2 hydrate growth. Finally, the Raman spectroscopy analysis indicated that a maximum of CO_2 dissolved was achieved within ~280 minutes. This study's findings on CO_2 uptake and the Raman analysis with seawater provide fundamental insights into CO_2 hydrate kinetics for carbon sequestration using gas hydrates.

Acknowledgments

PL acknowledges the funding support from the Agency of Science, Technology and Research (A*STAR) under the low carbon energy research (LCER) funding initiatives (project ID: U2102d2010).

173a CO_2 Adsorption and Utilization Using Silica Based Aerogel in Factory Chimney Environments.

Kyunghoon Min, Sangeun Shim

The current rise in atmospheric carbon dioxide (CO_2) levels is a serious global concern, primarily linked to the phenomenon of global warming. It stems largely from the combustion of fossil fuels and deforestation, leading to significant climate change worldwide. Global warming triggers events such as rising sea levels, melting glaciers and ice caps in polar and mountainous regions, and increased frequency of extreme weather events, posing serious threats to human health, safety, and the economy. Moreover, CO_2 remains in the atmosphere for long periods, meaning today's emissions affect future generations. Hence, efforts to reduce carbon emissions and promote alternative energy sources and sustainable industries are imperative.
Currently, significant efforts are being made to reduce and eliminate carbon dioxide emissions through ongoing research, with a particular focus on utilizing porous supports. Studies are investigating various methods, including impregnation and chemical grafting, to enhance the adsorption and utilization of carbon dioxide. Porous supports must withstand diverse environmental conditions, and research is underway to ensure their recyclability and other factors.

We have conducted research on utilizing a silica-based aerogel supported with PVP, which we developed, to capture CO_2 emissions from factory chimneys, one of the sources of carbon dioxide emissions. Employing a physical impregnation technique, we ensured that the aerogel could effectively adsorb CO_2 even at high temperatures while maintaining good recyclability. Additionally, to utilize captured CO_2, we introduced Zn^{2+} to facilitate the cycloaddition reaction of propylene oxide, an epoxide compound, within the aerogel support. This process aims to synthesize propylene carbonate as the product.

As a result, we measured CO_2 adsorption under various temperature conditions, notably observing a high adsorption rate of 10.2 mmol/g at 70 °C and 1 bar pressure. Moreover, the CO_2 selectivity in mixed gases was excellent. When employing these aerogel supports as catalysts, they sustained catalytic reaction for approximately 65 days at 70°C and demonstrated high purity of the resulting product.

This research provides a perspective on utilizing carbon dioxide emissions from factory chimneys by employing porous supports, specifically aerogels, to enable both adsorption and cycloaddition reactions via gas-phase processes.

173b The Role of Support on the Absorption Capacity of the Supported Ionic Liquid Membranes for CO_2 Capture

Sarang Ismail, Jarod Harris, Taylor Adams, Syed Ibrahim Gnani Peer Mohamed, Siamak Nejati, Mona Bavarian

In light of the current energy demand, our reliance on fossil fuels has increased exponentially. The environmental impact associated with this practice has led many to seek approaches that enable the capture of carbon emissions. Conventional carbon capture techniques, such as amine absorption, are employed to address the resulting carbon footprints. However, these methods have drawbacks in terms of operational and environmental efficacy, as well as concerns regarding toxicity and operational expenditures. To address these constraints, numerous researchers have directed their efforts toward identifying alternate approaches, with Ionic Liquids [IL] emerging as one of the most viable contenders. Due to their characteristics such as low vapor pressure, high thermal stability, customizable structure, low flammability, and high affinity for CO_2, ionic liquids (ILs) offer an excellent alternative to conventional carbon capture systems. Click or tap here to enter text. Recent studies have shown that nanoconfinement of ionic liquids (ILs) within polymeric domains has increased their CO_2 solubility. This suggests that ILs can be stabilized within the macromolecular matrix of the polymers and their non-ideal features can be utilized. It is crucial to comprehend the fundamental phenomena that drive the non-ideal behavior of ionic liquids (ILs) when they are integrated into polymeric domains. This study delves into assessing the influence of polymer molecular weight (MW) on the CO_2 sorption capacity of supported ionic liquid membranes (SILM). Quasi-solid-state films were produced by combining 1-Ethyl-3-Methylimidazolium bis(trifluoromethyl sulfonyl)imide and Polyvinylidene Fluoride (PVDF) with various MWs (180,000, 275,000, 352,000, 534,000, 1,000,000 g/mol) and Kynar. The CO_2 solubilities in these films were determined using the gravimetry method. The adsorption isotherms were measured concurrently with the heat of adsorption. The findings of our study demonstrate that the MW of PVDF plays a crucial role in determining the solubility of CO_2 in SILMs. This emphasizes the significance of the intermolecular interactions, polymer swelling, and PVDF polymorphism in influencing the absorption of CO_2 in SILM. Our research indicates that the mixture of PVDF(180K) and IL formed using a 1:1 ratio shows the highest potential for synthesizing SILMs. This mixture is capable of absorbing 3.1 moles of CO_2 per kg of IL. The value is threefold greater than the CO_2 sorption capacity of IL tested under the same conditions.

173c Experimental Study on the Sequestration of CO_2 through Mineral Carbonation of Ground Solidification Material

Heejun Kim, Hyun Sic Park, Jo Hong Kang, Sun-Yup Hwang, Hojun Song

Mineral carbonation, one of the CCU technologies, is known to have a practical CO_2 reduction effect in some countries with insufficient storage space. It is essential to secure raw materials and appropriately use carbonate after carbonation to realize mineral carbonation technology.

In this study, we attempted to immobilize CO_2 using ground solidification materials used in civil engineering construction as a raw material for mineral carbonation. Ground solidification materials generally contain a large amount of CaO, the raw material for pozzolanic reaction and ettringite formation. When applied to the ground, the solidification material is used together with water, and the hydrated CaO continuously hardens through a pozzolanic reaction to produce a stable, solidified body. Additionally, ettringite creates a more solidified body by evenly filling the space between soil particles in the ground.

To confirm the CaO content required for the reaction, component analysis of the solidification material, which is the raw material for carbonation, was conducted. In addition, the production of carbonate after carbonation was confirmed through XRD, XRF, and TGA analysis, and the amount of fixed CO_2 was calculated through this. To check the difference in strength depending on the carbonation ratio, a prototype was produced, and whether it met the standard strength (0.8 MPa or more, based on seven days of curing) was checked using a universal testing machine (UTM). This confirmed the possibility of future use in actual construction sites.

Keywords: Mineral Carbonation; Ground solidification material; Pozzolanic reaction; Ettringite; CCU.

Reference:

[1] Chen, Tiefeng, and Xiaojian Gao. Journal of CO_2 Utilization 34 (2019) 74-86.

[2] Siva, T., et al. Journal of Inorganic and Organometallic Polymers and Materials 27 (2017) 770-778.

[3] C. Han, Seong Ho, Yubin Jun, and Jae Hong Kim. Sustainable Materials and Technologies 35 (2023) e00559.

[4] Villagrán-Zaccardi, Yury Andrés, et al. Materials and Structures 50 (2017) 1-10.

173d Experimental Investigation of Microwave-Based Direct Air Capture Technology Using Zeolite 13X in a Fluidized Bed Reactor

Rahim Boylu, Mustafa Erguvan, Shahriar Amini

Direct Air Capture (DAC) technology has recently considered as an attractive technology for reduction of CO_2emission from the atmosphere. However, the technology requires some development in terms of capital cost, thermal energy demand, and regeneration time. Several studies have researched temperature swing adsorption method in order to improve relatively low energy demand of sorbent materials in the regeneration process [1-3], while some other investigations concluded that the sorbents commonly applied for CO_2 capture technologies have low conductivity in heat transfer from the sorbent to the captured CO_2 [4-6]. This heat transfer limitation negatively impacts the energy demand of the system, capital cost, CO_2 purity and CO_2 productivity. Even if relatively higher desorption temperature can lead to get rid of the heat transfer limitation, regeneration temperature is restricted by the stability of sorbents [7,8]. For this reason, microwave-based DAC application can be considered to make the regeneration possible under low temperature condition removing heat transfer limitation due to volumetric heating. etc.

In this purpose, this study investigates experimental microwave-based low temperature DAC using Zeolite 13X as a solid sorbent in a different sized-fluidized bed reactor size (ID 16, 26, and 41 mm). Dry compressed air in Tuscaloosa, Alabama has been applied as air source throughout the adsorption process. In the regeneration process, a mono-mode solid state microwave generator has been adjusted to preset temperature (ranging from 33 °C to 60 °C) and to initial power (switching from 5 W to 30 W). Various cycling experiments have been undertaken to identify regeneration performance in terms of adsorption capacity of the sorbent material in different cycles, energy consumption, regeneration time, desorption kinetics and CO_2 productivity. Additionally, since temperature homogeneity of the sorbent is challenging under microwave heating conditions, temperature distribution of Zeolite 13X with an average particle size of 217 microns has been measured using thermocouple, pyrometer, and IR camera during the regeneration stage. The results of this study prove that CO_2 regeneration is achievable at low temperature values under microwave heating in DAC application which also significantly contributes to reduction of energy demand of the system.

References

1. Kulkarni, A.R. and D.S. Sholl, Analysis of equilibrium-based TSA processes for direct capture of CO2 from air. Industrial & Engineering Chemistry Research, 2012. 51(25): p. 8631-8645.
2. Jang, G.G., et al., Process intensification of CO2 capture by low-aqueous solvent. Chemical Engineering Journal, 2021. 426: p. 131240.
3. Singh, P., W.P.M.V. Swaaij, and D.W.F. Brilman, Energy Efficient Solvents for CO2 Absorption from Flue Gas: Vapor Liquid Equilibrium and Pilot Plant Study. Energy Procedia, 2013. 37: p. 2021-2046.
4. Bos, M.J., et al., Study on transport phenomena and intrinsic kinetics for CO2 adsorption in solid amine sorbent. Chemical Engineering Journal, 2019. 377: p. 120374.
5. Wurzbacher, J.A., et al., Heat and mass transfer of temperature–vacuum swing desorption for CO2 capture from air. Chemical Engineering Journal, 2016. 283: p. 1329-1338.
6. Bos, M.J., et al., Evaluating Regeneration Options of Solid Amine Sorbent for CO2 Removal. Industrial & Engineering Chemistry Research, 2018. 57(32): p. 11141-11153.
7. Yu, Q., et al., Stability of a Benzyl Amine Based CO2 Capture Adsorbent in View of

Regeneration Strategies. Industrial & Engineering Chemistry Research, 2017. 56(12): p. 3259-3269.

8. Sonnleitner, E., G. Schöny, and H. Hofbauer, Assessment of zeolite 13X and Lewatit® VP OC 1065 for application in a continuous temperature swing adsorption process for biogas upgrading. Biomass Conversion and Biorefinery, 2018. 8(2): p. 379-395.

173e A Study on the Material Characteristics for Trace CO_2 Capture

Jo Hong Kang, Hyun Sic Park, Hojun Song

Carbon dioxide (CO_2) capture is becoming important both in environmental protection and in many areas related to energy security. it is essential to capture and separate traces and concentrated CO_2 to decrease the concentration of CO_2 in the atmosphere and to improve the quality of important commodities such as CH_4, H_2, and N_2. However, the technology involves a non-voluntary reaction that concentrates low-concentration CO_2 to high concentration, which poses significant challenges such as low technological maturity and high costs. To address these issues, improving the performance of low-concentration CO_2 capture technology is crucial, and for this purpose, the development of sorbents is essential.

In this study, various materials were utilized to capture trace amounts of CO_2 by considering parameters necessary for capturing trace amounts of CO_2, such as excellent CO_2 capture capacity at low pressure, high selectivity, and low heat of reaction, and a new material suitable for capturing trace amounts of CO2 was developed. The developed material was evaluated through various characterization techniques such as BET, XRD, and IR. The developed sorbent was evaluated for its CO2 sorption performance in a low-concentration CO2 atmosphere and its durability during long-term operation. In addition, its hydrothermal stability was confirmed by exposing it to real atmospheric conditions. Based on these results, the potential of the developed material as a sorbent for trace CO2 capture was confirmed, and it can be applied to processes under low-concentration CO2 conditions such as DAC, LNG power plants, and H2 PSA. In addition, it is expected to be utilized in various CO2 capture processes such as air purifiers and power plants.

Keywords: Carbon Dioxide capture, Traces CO2 removal, Selectivity, Heat of reaction, Characteristics

Reference:

[1] Belmabkhout, Y., Guillerm, V., & Eddaoudi, M. Chemical Engineering Journal 296 (2016) 386-397.

[2] Luukkonen, Aaro, Jere Elfving, and Eero Inkeri. Chemical Engineering Journal 471 (2023) 144525.

[3] Darunte, Lalit A., et al. ACS Sustainable Chemistry & Engineering 4.10 (2016) 5761-5768.

[4] Kumar, Amrit, et al. Angewandte Chemie International Edition 54.48 (2015) 14372-14377.

173f Sustainable and Feasible Carbon Capture and Utilization Pathways Towards Net-Zero Emissions

Ha-Jun Yoon, Tesfalem Atsbha, Taeksang Yoon, Dong-Kyoung Shin, Jihong An, Mohammadamin Zarei, Ali Cherif, Sangwon Suh, Chul-Jin Lee

To attain global goals for reducing greenhouse gas emissions, it is crucial to not only expand the use of renewable energy technologies and enhance efficiency but also recognize CCU as a key method for future CO_2 reduction. Despite achieving complete decarbonization of electricity production through widespread adoption of renewable energy, substantial CO_2 emissions from process-related activities in industries will persist. CCU offers a significant reduction potential for such emissions. Moreover, incorporating a portion of the captured CO_2 as a raw material in the chemical industry provides a sustainable approach rather than mere disposal. This utilization strategy also lessens the chemical industry's reliance on fossil fuels. Existing scientific literature showcases successful experiments demonstrating CO_2's feasibility as a raw material. However, not all CCU products have the capacity to efficiently utilize substantial amounts of CO_2 or be economically viable. Therefore, this current study concentrates on investigating CCU processes, warranting further exploration and comparison with conventional process routes based on carbon footprint, Techno-economic feasibility, and market potential by proposing novel methodology. In this study, a novel sustainable feasibility index (SFI) was developed to rank promising CCU processes based on techno-economic, life cycle assessments, and market size considerations. Eleven high-potential CCU products were investigated. Acetic acid, formic acid, and calcium formate were identified as the top-ranking CCU processes when compared with the SFI indicators for the base cases. Results from carbon to hydrogen mass ratio indicated decreasing the levelized cost of product (LCP) and global warming index (GWI) of CO_2 have a larger effect compared to decreasing the GWI of H_2 on overall carbon footprint of the CCU product. Future projections with respect to raw material sources indicate that the changes in CO_2 sources when compared to changes in the hydrogen and electricity sources, have the most significant effects on the CCU product LCP and GWI.

173g Evaluation of the Performance of CO_2 Absorbent Solutions using Bench Scale Membrane Contactor Systems

Hojun Song, Hyunji Lim, Gyeongmin Baek, Jieun Kim, Hyun Sic Park, Jo Hong Kang

Membrane contactors dramatically increase the specific surface area between gas and liquid compared to conventional packed towers, which can reduce the volume by up to a tenth. The small process volume, low weight and ease of scale-up of the membrane contactor-based CO_2 capture process make it suitable for small-scale applications such as biogas plants and CO_2 capture from ships. Hollow fiber membrane modules made of polypropylene are suitable for economical membrane contactor based CO_2 capture processes due to their low cost and hydrophobic nature. The CO_2 absorbent solution used in this technology must be able to reduce pore wetting of the membrane during long-term use. It should also have a good combination of properties such as fast CO_2 uptake, high CO_2 removal capacity and low regeneration heat of the absorbent solution.

In this study, experiments were carried out to select the optimum CO_2 absorbent solution using a 2 Nm^3/hr bench scale membrane contactor system and proprietary CO_2 absorbent solutions. The circulating flow rate of the CO_2 absorbent solution was controlled, and the CO_2 removal efficiency and regeneration heat of the absorbent were measured and compared according to the liquid-to-gas (L/G) ratio. In addition, to verify the membrane pore wetting characteristics of each CO_2 absorbent solution, the condensed absorbent solution in the knock-out drum was measured and compared. The surface tension and vapor pressure data of the CO_2 absorbent solution were further evaluated. In conclusion, it was found that the developed CO_2 absorbent solution has about 40 % lower regeneration heat and half the membrane pore wetting properties compared to the 30 wt% monoethanolamine (MEA) solution. Key factors for future scale-up were also discussed.

Keywords: Carbon Dioxide, Absorbent Solution, Membrane Contactor, Membrane Pore Wetting, Regeneration Heat of CO_2 Absorbent

Reference:

[1] I. Iliuta, F. Bougie, M. C. Iliuta, AIChE Journal 61 (2015) 955-971.

[2] H. Lim, K. Kim, H. S. Park, J.H. Kang, J. Park, H. Song, Journal of Industrial and Engineering Chemistry 122 (2023) 200-209.

[3] J. Kim, K. Kim, H. Lim, J. H. Kang, H. S. Park, J. Park, H. Song, Journal of Environmental Chemical Engineering 12 (2024) 112664.

[4] R. Ramezani, S. Mazinani, R. D. Felice, Reviews in Chemical Engineering 38 (2020) 273-299.

173h Modified Amine-Based Sorbent in Monolithic Structure for Direct Air Capture

Joo-Youp Lee, Dinabandhu Patra, Soumitra Payra, James Akinjide

A large portion of CO_2 emissions in ~5.2 gigatons (Gt) is released in relatively small quantities from distributed sources emitted each year in the U.S. Therefore, for such emissions, point source CO_2 capture is not feasible and direct air capture (DAC) is an indispensable part of a diversified portfolio of technologies to mitigate U.S. greenhouse gas emissions. Successful DAC technologies require the separation of high purity CO_2 (e.g., >95%) with high selectivity toward CO_2, low energy requirements, minimum chemical and thermal degradations, reliability, long lifetime, etc.

Adsorption-based CO_2 separation for DAC requires low pressure drop and low desorption energy. In this study, modified amine-based sorbent in a powdered form was washcoated onto a monolithic structure. Its performance was compared with that of unmodified amine-based sorbent on the same monolithic structure. Modified amine-based sorbent washcoated on a monolithic structure showed slightly lower adsorption capacities than unmodified sorbent on the same structure due to lower binding energy between adsorbed CO_2 molecules and modified amine sites. However, modified amine-based sorbent on the monolith showed higher CO_2 recovery than unmodified amine-based sorbent on the same monolithic structure under the same temperature vacuum swing adsorption conditions. The amount and purity of gases recovered from the monolith were analyzed using CO_2 analyzer and mass spectrometer.

Based on the performances of the small monolith blocks, the performances of large monolith blocks with unmodified and modified amine-based sorbents were also evaluated in a DAC system placed in an environmental chamber under various temperature, humidity and wind speed conditions simulating different climatic conditions. In this presentation, the performances of modified amine-based sorbent on monolithic structure will be compared with those of unmodified amine-based sorbent on the same structure. The performances of these two sorbents on the monolithic structure will be compared in terms of adsorption capacities, kinetics, desorption energy requirements, and stability.

173i The Hidden Chokepoints: Exploring Gas Diffusion in Codh/ACS Enzyme Complex Using Molecular Simulations.

Suman Samantray, Bojana Ginovska-Pangovska, Simone Raugei

The carbon monoxide dehydrogenase (CODH) and acetyl-CoA synthase (ACS) enzymatic complex catalyzes the (reversible) electrochemical conversion of CO_2 to CO and the subsequent condensation of CO with a methyl intermediate and coenzyme A to form acetyl coenzyme A. CO_2 reduction happens at the CODH C-cluster, and the CO product is transported and consumed at the ACS A-cluster. The bifunctionality of the complex is facilitated by a 140 Å-long gas channel for the transport of CO between the C- and A-clusters.

CODH and the CODH/ACS are central for atmospheric carbon monoxide oxidation and supporting microbial life. These enzymes are also the focus of recent biotechnological applications because of their potential replacement of two energy-intensive industrial processes: 1) the water–gas shift reaction, a widely used industrial process for H_2 or CO production, and 2) the Monsanto process used to produce acetic acid from methanol and CO. However, the effective and widespread use of CODH and CODH/ACS for industrial fields is limited by their poor activity under aerobic conditions. One strategy to design O_2-tolerant complexes is to prevent O_2 from entering the enzyme matrix.

In this work, we use atomistic free energy simulations to explore the gas diffusion through the CODH/ACS complex from M. thermoacetica and other design variants that favor uptake of CO_2 and O_2. We also explore how the transport mechanism is affected by conformational changes in the subunits of the enzyme caused due to transformation of cavities and gating mechanism of residues.

Research Interests:

"Peptoids" as a viable biomaterial in food packaging industry.

Antifreeze proteins (AFPs) stand as a paradigm-shifting innovation, that prevent freezing in organisms living in sub-zero environments by inhibiting ice crystallization. AFPs, such as Type I AFP from winter flounder and Type III AFP from ocean pout, have been instrumental in preventing ice crystallization in biological and industrial settings. This unique function has profound implications in cryopreservation, organ transplant technology, and notably in the food industry, where AFPs play a pivotal role in enhancing product longevity and quality.

Peptoids are synthetically engineered polymers, that could mimic natural AFPs with enhanced flexibility. The fundamental distinction between peptoids and peptides, while structurally similar, differ in the placement of their side chains. In peptides, side chains are attached to the alpha carbon atoms, whereas in peptoids, they are bonded to the nitrogen atoms of peptide backbone. Peptoids could be customized with targeted functional properties, to increase stability against enzymatic degradation, higher resilience under various physical conditions, and a reduced risk of immune responses, making them more suitable for industrial applications.

The research will address critical scientific questions: How do structural variations in peptoids influence their binding efficiency and antifreeze activity compared to natural AFPs? What thermodynamic and kinetic parameters are crucial in determining the antifreeze efficiency of peptoids? How can molecular dynamics simulations contribute to the design of optimized peptoid AFPs for industrial applications, particularly in food packaging?

Circular economy: Conversion of biomass-plastic waste to biochar-based adsorbents.

The crisis of plastic waste, particularly the accumulation of microplastics in the environment, presents a critical challenge for sustainable development. Addressing this issue requires innovative solutions that not only decompose plastic waste but also repurpose it into beneficial materials. Artificial metalloenzymes (ArMs) offer a targeted approach to catalyze the degradation of resistant plastic polymers, transforming them into value-added products such as biochar. Biochar, derived from the carbonization of organic matter, is recognized for its adsorptive properties, and is proposed mechanism to sequester pollutants.

Natural metalloenzymes like laccase and manganese peroxidase catalyze essential reactions in biological systems, including the degradation of complex molecules. Mimicking natural enzymes, artificial metalloenzymes can be engineered to target the sturdy chemical bonds in plastics such as polyethylene terephthalate (PET), high-density polyethylene (HDPE), and polystyrene (PS). Advanced molecular simulation techniques coupled with rational design and directed evolution can finely tune the active sites of metalloenzymes for specific substrate interaction, enhance catalytic rates, and provide stability against harsh industrial conditions.

The research aims to: Design artificial metalloenzymes with enhanced specificity for breaking down multiple types of plastics, focusing on the creation of intermediates destined for biochar conversion. Explore metabolic engineering of microorganisms to host these ArMs, evaluate the life-cycle of the enzymatic byproducts, and upscale the entire process for industrial use. Successful implementation would represent a dual achievement: mitigating plastic pollution and harnessing the resulting biochar for environmental adsorption applications, thereby fostering a circular economy in waste management.

PFAS-mediated dysregulation of biomolecular condensates.

The pervasive presence of per- and polyfluoroalkyl substances (PFAS) in the environment, coupled with their long biological half-life, has emerged as a concern for public health. PFAS compounds like PFOA and PFOS, notorious for their toxicity, exposure to it is linked to a multitude of health risks, and there is a critical need to understand their interaction with cellular components at a molecular level. This research focuses on the impact of PFAS on biomolecular condensates, which are crucial for various cellular processes, including RNA metabolism, signal transduction, and the stress response.

Biomolecular condensates, formed via liquid-liquid phase separation, are highly dynamic and sensitive to changes in their microenvironment. The physicochemical properties of PFAS, such as their hydrophobicity and stability, make them potential disruptors of the condensates' structural integrity. PFAS affect biomolecular condensates at the molecular level, this research aims to utilize molecular simulations to bridge the gap between environmental exposure and cellular dysfunction, offering a foundational perspective on the health implications of PFAS. This work has the potential to inform risk assessments, shape regulatory decisions, and inspire the development of therapeutic interventions to counteract the detrimental effects of PFAS on human health.

Teaching Interests:

The beauty of academic life is that it gives an opportunity to learn, teach, and work with young, bright, and dynamic students, who are the future of our society. Through the influences of the great teachers, I am at this stage to pursue academic life as a part of devoted teaching and research to serve our society. In the following, I describe in detail experiences, teaching philosophy, and plan for teaching.

Teaching experience: I have had a privilege of being TA for the computational biotechnology practical courses of MSc students at University of Galway, Ireland and RWTH Aachen University, Germany. In addition, I co-mentored theses of BSc, MSc students, and students.

Teaching Philosophy: In my opinion, the most important aspect of teaching is that we learn, teach, work, and develop together for the welfare and service of our society. I believe that we can learn best by 'doing'. The fundamental aspect of teaching is igniting the mind of students with the curiosity and skills necessary to develop questions as well as guide them one-step further to motivate, promote and develop skills and scientific processes necessary to investigate the answers to those questions. To facilitate this, I will keep my classes interactive and open for discussion. Besides, I will take the opportunity to discuss teaching, learning, training, and research with students, friends, colleagues, senior professors, and experts for the betterment of my service to the institute and society.

Teaching Plan: I am interested in teaching the core and professional courses of Biochemical Engineering. In the core courses, I am in the teaching of Protein thermodynamics, Statistical mechanics, Chemical kinetics, and Chemical reaction Engineering. In the professional courses, I would like to teach Bioinformatics, Structural Biology, and Biophysics. Besides, I have the interest to design courses, e.g., Bioeconomy, Protein Engineering.

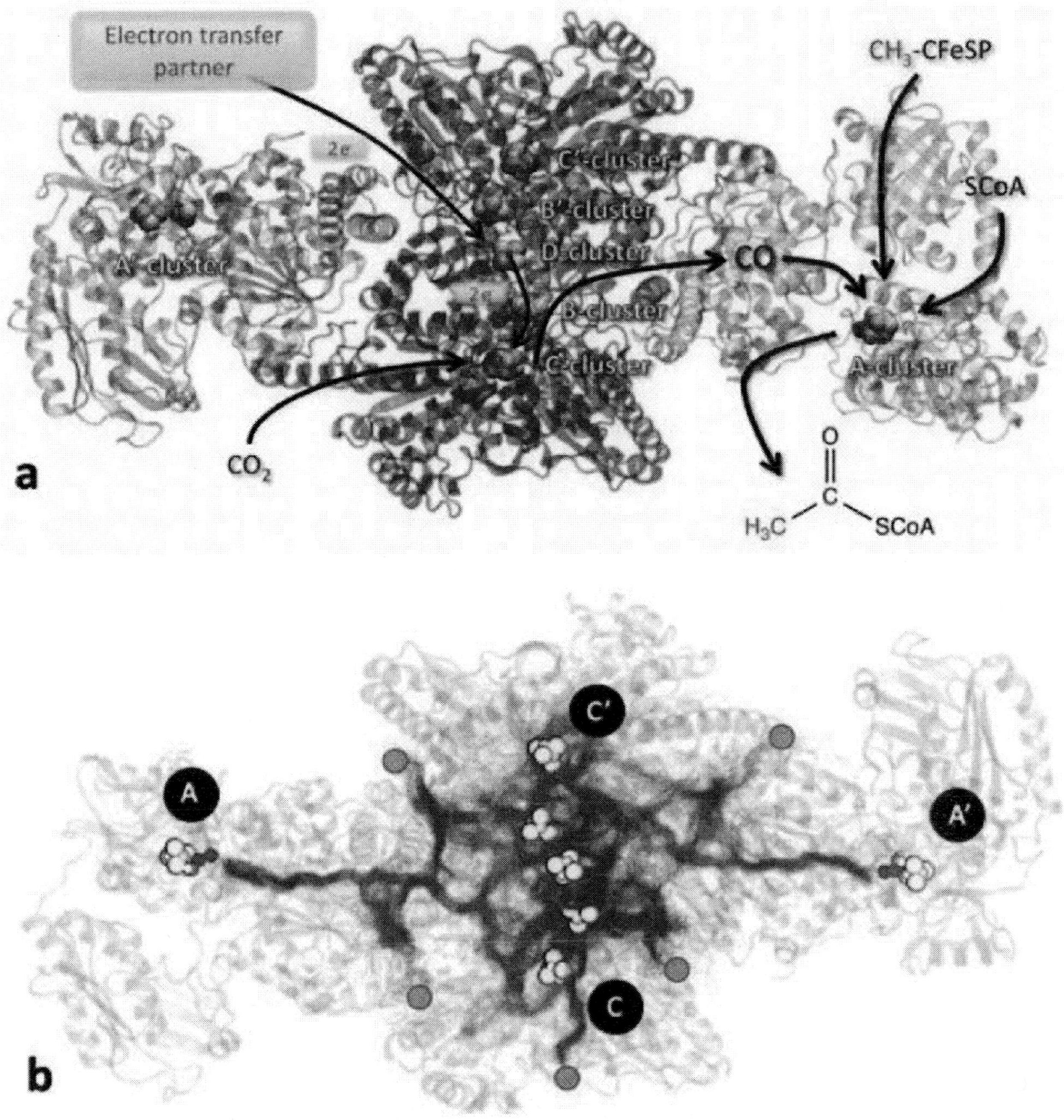

Figure 1. a) Schematic representation of the CO_2/CO processes catalyzed by the CODH/ACS complex from *M. thermoacetica*, with the active metalloclusters A- and C- labelled in cyan. b) A composite view of gas channels (dark gray) in CODH/ACS discovered by long-timescale molecular dynamics simulation.

173j Application of Fluidized Bed Reactors for CO2 Hydrogenation for the Production of Hydrocarbons

Hyungseok Nam, Young Min Yoon, Eun-Ho Jang, Doyeon Lee, Byungwook Hwang, Sun-mi Hwang

The major sources of CO2 emissions in the oil refining process include atmospheric and vacuum distillation processes, reforming processes, fluid catalytic cracking (FCC) processes, and hydrogen production processes. Particularly, the off-gas generated after the hydrogen production process contains about 24 vol% of low-purity CO2 along with H2 and CO, and it is the largest source of CO2 emissions in the refining process. Technologies for directly converting low-purity CO2 into hydrocarbons are being researched and developed. The reaction mechanism involves converting CO2 to CO via the reverse water-gas shift (rWGS) reaction, and then converting CO and H2 into hydrocarbons via the Fischer-Tropsch (FT) synthesis. Iron-based catalysts are most reported for satisfying both reactions simultaneously, and there are ongoing studies on improving conversion rates by adding alkali metal-based co-catalysts. However, most studies have been conducted in fixed-bed reactors, and there are problems such as catalyst degradation due to the exothermic nature of the FT reaction, which leads to a decrease in catalyst activity. Thus, research is needed to control the heat and solve the problem of local thermal runaway. In this study, a lab-scale bubbling fluidized bed reactor (BFB) was utilized for its good response to exothermic reactions and ability to maintain a constant operating temperature. Basic experiments were conducted to understand the heat transfer and CO2 conversion reaction characteristics. Heat transfer coefficients were derived through the cooling tubes installed in the reactor, and the optimal heat transfer coefficients were determined based on various fluidization velocities (0.725bar), and temperatures (200~400°C). Based on this, foundational research on CO2 conversion was conducted.

173k Recovery of NH_3 from Livestock Manure for Renewable, Net-Zero Fuel

Ken Tasaki

The nitrogen (N) fertilizer consumption in the U.S. has quadrupled since 1960. This trend has resulted in two consequences. One is the excess N in the soil, which is prone to run off to waterways, causing eutrophication and groundwater contamination. Another is greenhouse gas (GHG) emissions associated with nitrogen. Agriculture is responsible for 75% of the total U.S. N_2O emissions in 2021. 95% of agricultural N_2O emissions are due to soil management, including fertilizer applications. N_2O emissions are 265 times as potent as CO_2 emissions as GHG emissions, accounting for 58% of the total GHG emissions from agricultural activities in the U.S. Every MT of NH_3 generates 2.6 MT of CO_{2eq} emissions by the Haber-Bosch process. Currently, the U.S. consumes 14 MMT of NH_3, most of which is used as fertilizers. This corresponds to 36.4 MMT of CO_{2eq} emissions. In addition, applications of N fertilizers to the soil generate N_2O emissions, which is 5.6 % of the total CO_{2eq} emissions in the U.S., equivalent to 353 MMT of CO_{2eq} emissions. Together, the consumption of N fertilizers generates 389 MMT of the life-cycle assessment (LCA)-based CO_{2eq} emissions in the U.S. Studies show that 30 ~ 50 % of those N_2O emissions from soil management originate from applications of animal manure.

Recently, Green NH_3 or H_2 have been receiving strong attention as net-zero emission energy sources. However, both technologies use water electrolysis, which consumes a large amount of renewable electricity. Green NH_3 will require dedicated electricity with a capacity that would be 4.6 times the current global renewable energy production if the goal of the net-zero emission society is to be achieved by 2050. Further, the recent Oxford Report estimates the price of Green NH_3 to be 3.7 times more expensive than Gray NH_3 based on natural gas due to the high cost of electricity for water electrolysis. However, there is a vast, untapped resource for N in the livestock industry. According to one estimate, there were 6 MMT of N as a nutrient in manure discharged from livestock animals in the U.S. This number is more than half the total N fertilizer consumption in the U.S. and almost 90% of the N fertilizers the U.S. imports every year. This paper presents a novel N recovery process with very low LCA-based CO_{2eq} emissions and LCA-based energy consumptions. The recovered NH_3 can be used as renewable N-fertilizers, renewable net zero-carbon fuels, or H_2 carriers. H_2 can be produced by reforming NH_3.

The stripping/scrubbing process is a common process for NH_3 removal and recovery at industrial facilities. However, its CAPEX/OPEX still need to be lower for livestock farmers to adopt the technology widely. Also, the N concentration in the recovered solution is low, ~ 1 N% which is not high enough for efficient NH_3 condensation to produce liquid NH_3. It also requires high water consumption. More importantly, the current stripping/scrubbing design has limited mass-transfer interface areas, resulting in slow mass-transfer kinetics. Mass transfer is the key to designing the NH_3 recovery process in a commercially viable way and scale. Mass transfer by stripping/scrubbing has been extensively studied experimentally and theoretically. Still, the mass transfer-based NH_3 recovery is an unexplored field. For example, there has yet to be a university-accepted valid correlation between the liquid-to-gas (L-G) or the gas-to-liquid (G-L) mass-transfer rates and the design/operating parameters of a wastewater treatment plant. We believe that a thorough understanding of the relationships between the mass-transfer rate constants and the controllable design/operating parameters in both NH_3 stripping and absorption will help advance a fundamental scientific and engineering understanding of an L-G and G-L multiphase and multi-component mass transfer coupled with chemical reactions, a common phenomenon in many chemical and biochemical processes. Our system has a multi-component mass transfer; when NH_3 is stripped by air, it is mixed with N_2 and O_2 gases. Hence, the G-L transfer involves a mass transfer of the NH_3 gas mixed with N_2 and O_2 gases. Our mass transfer is coupled with chemical reactions because the process uses chemical reactions to facilitate the NH_3 stripping and absorption, especially when stoichiometric chemicals are used.

Accordingly, we first set out to develop a new mass transfer kinetic model (Model I): a consecutive L-GG-L mass transfer kinetic model which has rarely been reported in the literature: the L-G mass transfer for the NH_3 stripping and the G-L for the NH_3 absorption.

Then, we applied the model to our NH_3 recovery process. The model guided us to adopt column systems for their larger interfacial areas and their lower maintenance cost than those of conventional tower systems. According to Model I, the concentration in the absorption tank is expressed as follows:

(1)

where , , , , refer to the concentration in the absorption tank at the time, *t*, the absorption recovery yield in %, the stripping removal yield in %, the concentration in the stripping column in the beginning, and the mass transfer rate per unit volume for the absorption, respectively. The subscripts "*a*" and "*s*" refer to the absorption and the stripping, respectively. The mass transfer rate per unit volume for the stripping, , is embedded in . Hence, eq 1 combines the L-G and G-L mass transfers for the first time. Based on the model, we devised the ranges of the operation parameters to maximize both L-G and G-L mass transfers. Hence, our technology is a model-based recovery process. The operation parameters include the air flow rate for aeration, the orifice diameter of the air diffusers, the column design, the acid/base concentrations, and others.

In addition, we have developed a theoretical kinetic model (Model II) to evaluate the mass-transfer coefficient, , by applying the transition state theory to the mass transfer resistance at the interface, which is equivalent to . The L-G mass transfer coefficient, , has been widely investigated for a variety of gases and chemicals. Yet, the rationale for the dependence on has been scarce in the literature. As to the G-L coefficient, , very few data are available for NH_3 in the literature. With Model II, we were able to calculate the energy barrier for using the density functional theory. Then, we used Model II to investigate the dependence of on the concentration of either base (the NH_3 stripping) or acid (the NH_3 absorption) and gained a better understanding of how the operating parameters affect at the atomistic level.

Through repeated experiments under different operation parameters, we were able to establish the relationship between the mass transfer rate and the operation parameters. Using the relationship, we found the optimized condition to maximize the L-G or G-L NH_3 mass transfers. We employed chemical reactions to facilitate the NH_3 stripping and absorption. One example is the following reactions:

$Ca(OH)_2 + 2 \rightarrow 2NH_3 + Ca^{2+} + H_2O$ (2)

$2NH_3 + H_2SO_4 \rightarrow (NH_4)_2SO_4$ (3)

The stoichiometric amounts of the chemicals produce a 20 N% solution as the final product, which is high enough for the N condensation. Under the optimum conditions, we achieved the L-G mass transfer rate of 1,274.5 mg L^{-1} hr^{-1}, quadrupling some of the highest reported for the NH_3 stripping in the literature. As to the G-L mass transfer rate of 300 mg L^{-1} hr^{-1}, it is still higher than 250 mg L^{-1} hr^{-1} in a previous study. Both and were more than 90%. Our estimate shows that the LCA-based energy consumption for the recovery of 1 kg of NH_3 by our process is 6 kWh kg-NH_3^{-1} which compares with 11.3 and 13.7 kWh kg-NH_3^{-1} for Gray and Green NH_3, respectively. As to the LCA-based CO_{2eq} emissions, our proposed process using eqs 2 and 3 is estimated to generate 0.74 kg CO_{2eq} kg-NH_3^{-1}, which compares with 11 and 1.35 ~3.8 kg CO_{2eq} kg-NH_3^{-1} for Gray and Green NH_3, respectively. Our new process consumes about half the energy and generates less CO_{2eq} emissions than Gary or Green NH_3. Our cost analysis suggests that the cost of producing 1 MT of NH_3 is \$1,258, \$2,427, and €2,400 by our process, based on the current price of $(NH_4)_2SO_4$ and the estimate for the stripping/scrubbing process from literature, respectively. Once the N fertilizer is produced by eq 3, NH_3 can be recovered by conventional evaporation and liquefaction, both of which are routine processes. For the transportation of NH_3, infrastructure, such as storage and pipelines, already exist. As a demonstration, we will show the liquid NH_3 obtained by boiling a $(NH_4)_2SO_4$ solution and condensing the NH_3 gas under -30 °C.

The long-distance cargo shipping industry is currently pilot-testing NH_3 for bunker fuel, which is expected to be the next generation of fuel. However, the currently high cost of Green NH_3 prevents the industry from widely spreading its application. Our process, if successful, would promote a faster adaptation of renewable NH_3 fuel in this industry and beyond.

173m A Novel Lstm-CNN Framework Forecasts India's Air Quality Using Historical Data, Outperforming Traditional Methods. Insights Aid Tailored Pollution Mitigation.

Swetha .

Air quality deterioration presents significant challenges in India, necessitating effective assessment and prediction methods. This study proposes integrating memory-based learning techniques to comprehensively evaluate air quality parameters across Indian regions. Utilizing LSTM and CNN, our framework effectively captures temporal and spatial dependencies using historical data. Validation against diverse pollutant datasets, including PM2.5, PM10, NO2, SO2, and O3, demonstrates its superiority over traditional methods. Furthermore, it explores the intricate relationship between meteorological factors and air quality, offering insights for targeted pollution mitigation. Utilizing data from monitoring stations, satellite imagery, and various sources, advanced analytics and machine learning algorithms are employed to develop a predictive model.Air quality deterioration presents significant challenges in India, necessitating effective assessment and prediction methods. This study proposes integrating memory-based learning techniques to comprehensively evaluate air quality parameters across Indian regions. Utilizing LSTM and CNN, our framework effectively captures temporal and spatial dependencies using historical data. Validation against diverse pollutant datasets, including PM2.5, PM10, NO2, SO2, and O3, demonstrates its superiority over traditional methods. Furthermore, it explores the intricate relationship between meteorological factors and air quality, offering insights for targeted pollution mitigation. Utilizing data from monitoring stations, satellite imagery, and various sources, advanced analytics and machine learning algorithms are employed to develop a predictive model. A novel framework merging LSTM and CNN effectively forecasts India's air quality using historical data. Validated against diverse pollutant datasets, it outperforms traditional methods. Insights into meteorological influences aid tailored pollution mitigation. Advanced analytics, including machine learning, utilize varied data sources to construct predictive models. Historical pollution and weather data drive training for regional pollution forecasts, promising targeted interventions for improved air quality management.In response to India's pressing air quality challenges, a pioneering approach intertwining long short-term memory (LSTM) networks with convolutional neural networks (CNN) has emerged. This innovative framework utilizes extensive historical air quality data to effectively capture both temporal and spatial dependencies. Rigorous validation against comprehensive multi-pollutant datasets, encompassing pollutants such as PM2.5, PM10, NO2, SO2, and O3, underscores its superior predictive capability over traditional methods. Beyond prediction, the framework unravels the complex relationship between meteorological variables and air quality variations, furnishing critical insights for tailored pollution mitigation strategies. Drawing from diverse data sources including monitoring stations and satellite imagery, advanced analytics techniques, particularly machine learning algorithms, are harnessed to construct a predictive model. Historical pollution records and meteorological parameters drive model training, facilitating forecasts of pollution levels across diverse Indian regions within specified timeframes. This holistic approach promises to inform targeted interventions for sustainable air quality management in India. effective evaluation and forecasting methods. Leveraging historical air quality data spanning several years, a novel framework is developed, combining long short-term memory (LSTM) networks with convolutional neural networks (CNN) to capture temporal and spatial dependencies effectively.through meticulous validation using comprehensive multi-pollutant datasets covering various pollutants such as particulate matter (PM2.5, PM10), nitrogen dioxide (NO2), sulfur dioxide (SO2), and ozone (O3), we demonstrate the superior predictive performance of our proposed framework compared to conventional forecasting methods. Moreover, our analysis goes beyond mere prediction by delving into the intricate interplay between meteorological factors and air quality fluctuations, offering invaluable insights for the formulation and implementation of targeted pollution mitigation strategies Utilizing data from monitoring stations, satellite imagery, and various sources, advanced data analytics techniques, including machine learning algorithms, are employed to develop a predictive model. Historical pollution data and meteorological parameters serve as inputs to train the model, aiming to forecast pollution levels across different Indian regions over specific timeframes.

Air quality deterioration presents significant challenges in India, necessitating effective assessment and prediction methods. This study proposes integrating memory-based learning techniques to comprehensively evaluate air quality parameters across Indian regions. Utilizing LSTM and CNN, our framework effectively captures temporal and spatial dependencies using historical data. Validation against diverse pollutant datasets, including PM2.5, PM10, NO2, SO2, and O3, demonstrates its superiority over traditional methods. Furthermore, it explores the intricate relationship between meteorological factors and air quality, offering insights for targeted pollution mitigation. Utilizing data from monitoring stations, satellite imagery, and various sources, advanced analytics and machine learning algorithms are employed to develop a predictive model. Historical pollution and meteorological data drive model training to forecast pollution levels regionally. The model's potential applications include informing policymakers, urban planners, and environmental agencies for implementing effective pollution control measures and optimizing resource allocation to mitigate air pollution's adverse effects.

Air quality deterioration presents significant challenges in India, necessitating effective assessment and prediction methods. This study proposes integrating memory-based learning techniques to comprehensively evaluate air quality parameters across Indian regions. Utilizing LSTM and CNN, our framework effectively captures temporal and spatial dependencies using historical data. Validation against diverse pollutant datasets, including PM2.5, PM10, NO2, SO2, and O3, demonstrates its superiority over traditional methods. Furthermore, it explores the intricate relationship between meteorological factors and air quality, offering insights for targeted pollution mitigation. Utilizing data from monitoring stations, satellite imagery, and various sources, advanced analytics and machine learning algorithms are employed to develop a predictive model. Historical pollution and meteorological data drive model training to forecast pollution levels regionally. The model's potential applications include informing policymakers, urban planners, and environmental agencies for implementing effective pollution control measures and optimizing resource allocation to mitigate air pollution's adverse effects.

Air quality deterioration presents significant challenges in India, necessitating effective assessment and prediction methods. This study proposes integrating memory-based learning techniques to comprehensively evaluate air quality parameters across Indian regions. Utilizing LSTM and CNN, our framework effectively captures temporal and spatial dependencies using historical data. Validation against diverse pollutant datasets, including PM2.5, PM10, NO2, SO2, and O3, demonstrates its superiority over traditional methods. Furthermore, it explores the intricate relationship between meteorological factors and air quality, offering insights for targeted pollution mitigation. Utilizing data from monitoring stations, satellite imagery, and various sources, advanced analytics and machine learning algorithms are employed to develop a predictive model. Historical pollution and meteorological data drive model training to forecast pollution levels regionally. The model's potential applications include informing policymakers, urban planners, and environmental agencies for implementing effective pollution control measures and optimizing resource allocation to mitigate air pollution's adverse effects.

173n Seasonal Measurement of Greenhouse Gas Emissions from the Energy Capital of California

Zhongzhe Liu, Chengwei Lei, Liaosha Song, Huaiyu Zhang, Calvin Uclaray

Oil and gas industries are the core stone of the economy in the Central Valley of California. Kern County produces about 70% of the state's oil and more than 90% of its natural gas, making it the energy capital of California. In the meantime, a considerable amount of greenhouse gas emissions such as CH_4 is generated during oil and gas drilling activities and from idle oil and gas wells, which significantly contributes to the bad air quality and worsen the climate change. In particular, many idle oil and gas wells in the county were leaking explosive levels of CH_4, which poses threats to nearby residential communities and aggravates greenhouse effect. Due to the increasing pressures on the environment and California's ambitious goal of going carbon neutral by 2045, this study was aimed to obtain the preliminary emission data (i.e., CH_4 and CO_2 trace concentrations) and make emissions visible on the map to arouse the public awareness of both air quality and climate change in the energy capital of California. The first objective of this study was to determine the regions of oil and gas wells for the emission measurement. According to San Joaquin Geological Society, some productive oil fields in the county were chosen such as the Kern River oil field. The second objective of this study was to measure the trace CH_4 and CO_2 concentrations in different seasons from a selected oil field. A vehicle equipped with a CH_4/CO_2 trace gas analyzer was used to measure the real-time concentrations of CH_4 and CO_2. The vehicle moved on the roads within the predetermined route at a stable speed for data collection including the GPS data. The third objective was to integrate emission data and geographic data to visualize emissions on the map. The preliminary study showed that the CH_4 concentrations around the selected oil field were 2,500 ppb on average, which was slightly higher compared to the CH_4 concentration in the atmosphere of ~2,000 ppb. In some areas of the selected oil field, the CH_4 concentration was very high (e.g., 17,000 ppb) but still way below the explosion limit of 50,000 ppm in the air. A more frequent monitoring is needed in higher-concentration emission areas. Also, the measurement of greenhouse gas emissions from other oil fields in Kern County needs to be conducted in the future for comparison.

173o Chloride and Ammonia Removal By Dry Deposition during an Accidental Release: Considering the Importance of the Boundary Layer Resistance

Thomas Spicer III

Many toxic industrial chemicals (TICs) such as chlorine and ammonia are very reactive with commonly encountered environmental materials. The amount of TIC material reacted in the environment is an important factor in determining the impact of a release. Typically characterized with a dry deposition rate in atmospheric dispersion models, the reaction rate has predominately been studied in field scale experiments at concentration levels typical for air pollution (ppm levels). When considering the removal of chlorine or ammonia by (dry) deposition in an episodic release, gas phase concentrations will be much higher than those associated with air pollution, and consequently, reaction rates of chlorine and ammonia with local vegetation, soil, and built-environment materials are expected to be much higher. In this work, standard engineering methods characterizing the rate of mass transfer from a fluid to a surface in conjunction with reaction at that surface are applied, and analysis of experimental data showed that reaction of chlorine and ammonia reach a limiting maximum. This study developed a model to account for boundary layer resistance and the impact of maximum deposition in a framework that could be incorporated in atmospheric dispersion models. The Controlled Environment Reactivity Test (CERT) apparatus was designed to expose selected environmental materials to chlorine and (separately) ammonia with initial concentration of (nominally) 1000 ppm. The gas phase concentration was measured as a function of time to determine relevant modeling parameters. The apparatus was built to control the velocity of the gas/air mixture over the experimental substrates with turbulence levels that are comparable to the atmosphere. Model parameters were found for the environmental materials tested. Ammonia proved much more reactive than chlorine, and consequently, the characterization of the boundary layer resistance proved important for the data analysis.

173p *Automatic Chemical Reaction Mechanism Reduction Using Artificial Intelligence: Applications in Atmospheric Chemistry*

Arijit Chakraborty, Forwood Wiser, Siddhartha Sen, V. Faye McNeill, Venkat Venkatasubramanian

The objective of chemical reaction mechanism reduction is to obtain a condensed reaction mechanism that is a fairly accurate surrogate of the complete reaction mechanism and is considerably more memory and computationally efficient for subsequent tasks, such as real-time concentration prediction. When formulated as a search problem, the search space of potential candidate mechanisms could be larger than those of games such as Chess and Go combined. In this work, we present an artificially intelligent (AI) automated mechanism reduction of the gas-phase isoprene oxidation mechanism, termed Genetic Algorithmic - Automated Model Reduction (GA-AMORE). We highlight the efficacy of our approach in a well-established case study: atmospheric gas-phase isoprene oxidation. Formulated as a search problem through an astronomically large search space, we explore the same under mechanistic constraints, and subject matter expert rules, thus highlighting the benefits accrued from such a hybrid AI approach. We contrast the resultant mechanism(s) obtained to state-of-the-art manually tuned mechanisms of comparable size. Further, we obtain the optimal stoichiometric coefficients, and rate parameters, of the reduced reaction mechanisms using a derivative-free optimization strategy that significantly improves its accuracy. We posit that our approach can be extended to other mechanism reduction case studies by mere modification of the domain-specific rules, while keeping the underlying algorithm unchanged.

173q Photolytic Decomposition of Perfluorooctanesulfonic Acid (PFOS) By Composite Nanofibers of Fe(III)/PVDF Under UVC Light

Dipendu Saha, Savannah Steger, Christina Mccullough

Per- and polyfluoroalkyl substances (PFAS) are a general classification of over 3000 unique variations of various fluorinated alkyl compounds. PFAS are found in a wide variety of everyday industrial and household products. These substances are man-made and do not occur naturally in the environment. In North America, no specific guidelines have been put into place for the disposal of PFAS-based materials, and because of this, at the end of their lifecycle, they are typically discarded into landfills. As these materials gradually break down, they are released into landfill leachate, and as these materials travel through the rainwater into rivers and streams, they eventually enter and accumulate in various water sources, including those from which drinking water is obtained. As this PFAS-contaminated water is consumed, a bioaccumulation can occur which can lead to various types of serious illnesses and ailments. Among the over 3000 variations of PFAS molecules, perfluorooctane sulfonic acid (PFOS) is one of the most prevalent and difficult to remove. Due to the potential health risks that an accumulation of PFOS in drinking water can pose, along with the general environmental risks, a water treatment process designed to remove or decrease the concentration of PFOS in drinking water is a topic of severe importance and a crucial area of research.

In this research, an experimental procedure has been designed and evaluated in which Perfluorooctanesulfonic acid (PFOS) has been photolytically degraded. Nanofiber composite with Fe(III) and polyvinylidene fluoride (PVDF) was fabricated by electrospinning and characterized with TGA, XRD, SEM, and AFM. The nanofiber composite successfully degraded Perfluorooctanesulfonic acid (PFOS) in aqueous medium and in the presence of UVC light. It was observed that the presence of hydrogen peroxide increases the rate of the reaction. It is suggested that PFOS binds with Fe(III) and a free radical was generated associated with PFOS molecule. The free radical is further degraded to smaller molecular fragments in the presence of UVC light. The study investigated the rate of PFOS degradation as a function of time, concentration of PFOS and hydrogen peroxide, and number of composite mats. The results demonstrated Fe(III)/PVDF nanofiber composite can potentially be used for PFAS removal.

173r Conversion of Waste Plastic to Oils for Feed to High-Pressure Gasifiers

Angela Darko, Ignacio Preciado, Kevin Whitty

The focus of this work is to understand the properties of oils produced from waste plastics and make plastic oils specifically tailored for gasification. The plastic oil will be one of the components of a slurry that also consists of biomass pyrolysis liquids and ground coal waste; this blend will be gasified for the production of hydrogen or other high-value products.

Polystyrene (PS), Polypropylene (PP), Low-Density Polyethylene (LDPE), High-Density Polyethylene (HDPE), and mixtures of these plastic pellets are subjected to thermal pyrolysis in a two-stage fixed-bed reactor (batch system) to create improved plastic oil that flows easily at room temperature for gasifier feed. The reactor is operated over a range of conditions (including various temperatures and residence times) to understand relative yields and compositions of liquid, gas, and char, as well as the viscosity characteristics of the resulting liquid. The goal is to identify conditions that will produce suitable plastic oil for gasification with minimal processing intensity.

For each test, 100 g of plastic pellets were loaded inside a container that was placed into the first section of the reactor. As pyrolysis vapors evolve from the plastic bed, they are carried into the second section of the reactor by the nitrogen carrier gas with a flow rate of 200 ml/min. Reactor sections 1 and 2 can be heated at different temperatures (400 to 600ºC), then, the condensable plastic pyrolysis volatiles are collected as plastic oil in a series of chillers, which are chilled to

-15ºC. Two acrylic tubes in series filled with activated carbon and silica gel, respectively, are used to clean the gas of any remaining oil and moisture before passing through a micro-GC, which allows the identification of resulting gaseous species such as H_2, CO, and light hydrocarbons (C_1 through C_3).

Figures 1 and 2 summarize the preliminary experimental results obtained from the pyrolysis process in the two-stage fixed bed reactor.

1. a) Liquid Yield: Represents the mass % of the material loaded that was collected as liquid in the chilled jars.
2. b) Char Yield: Represents the mass % of the material loaded in the reactor that stayed as carbonaceous/char-type material.
3. c) Gaseous Yield: represents the expected mass % of the material loaded that should be converted to gaseous species, these numbers are obtained by difference. % of gaseous products yield = 100% minus % reported in a+b.

Liquid products: PS produces large amounts of liquid products (more than 95% of the mass loaded) at both 500ºC and 600ºC (a slight yield reduction is observed at 600ºC vs 500ºC). The liquid yield for PP is considerably larger at 500ºC than at 600ºC, but the resulting liquid product at 500ºC is very thick and flows with difficulty at room temperature. About 70% mass of the loaded plastics (LDPE and HDPE) is collected as liquid material when pyrolyzed at 600ºC, but they are thick fluids that flow with difficulty at room temperature. These two plastics (LDPE and HDPE) were not tested at 500ºC because they would have created even thicker materials that could have clogged some components of the experimental system.

The plastic oils produced from PS, PP, and the mixture of PS-PP-HDPE (33.3% each) are liquids and flow easily. The presence of PS and PP in the mixture helps dissolve the thick product created from HDPE.

Solid products: Small amounts of char-type material are created from all plastics (around 1% or less of the mass loaded). LDPE and HDPE produce the higher amounts, followed by PS and PP.

Gaseous products: The expected yield % of gaseous species needed to close the mass balance of the test are shown in figures 1 and 2 (gaseous products, yield %), but the gaseous species detected by the micro-GC (hydrogen, methane, ethylene, ethane, propylene, and propane), are only a portion of the total gaseous species.

Other mixtures/plastic ratios will be tested to better understand the influence of PS and PP on polyethylene plastic oils and rheology measurements will be conducted to determine the viscosities, cloud point, and pour point of the plastic oils.

Other parameters to be evaluated include changing the residence time of the gases in the pyrolysis reaction zone (by changing the N_2 flow rate), the presence of contaminants such as oil, paper, soil, and bio-solids in the plastic pellets and see their effect on plastic oil characteristics.

Figure 1: Summary Yield of Pyrolyzed Plastics at 500°C

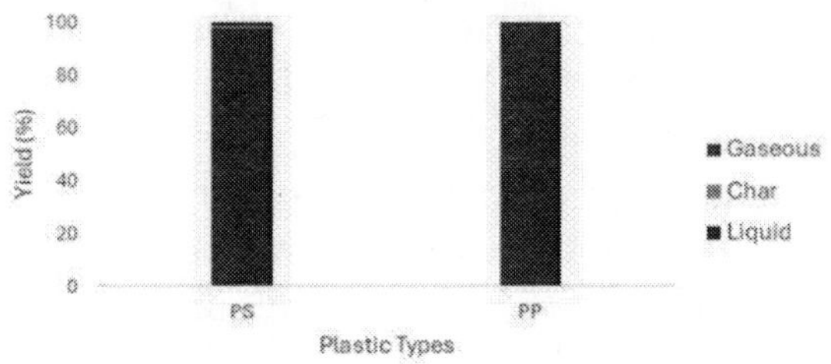

Figure 2: Summary Yield of Pyrolyzed Plastics at 600°C

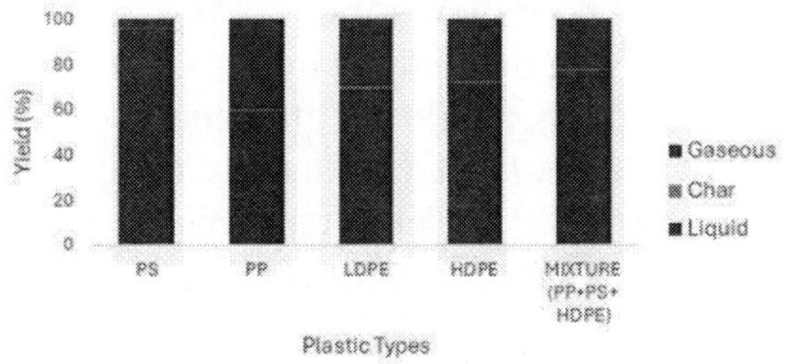

173s Microplastics in Wastewater Treatment Facilities: Fate and Transport in the Cookeville Wastewater Treatment Plant and Literature Review

Sahera Abumariam, Shafieh Karami, Phuong Tran, Pedro Arce

Microplastics (MPs) are considered as plastic particles less than 5 mm in size, and like other anthropogenic contaminants, they originate from a variety of land-based sources and eventually find their way into municipal wastewater treatment plants. These facilities are widely described as a link between pollutants and natural environments (Turan et. al 2021). More than 80% of MPs enter the ocean from land, primarily from terrestrial activities such as fishing, aquaculture, and coastal tourism, while less than 20% come from the sea, demonstrating their ability to travel far due to their light weight, durability, and buoyancy (Yee et. al 2021). MP particles are able to put both mankind and the environment in danger. Several studies have shown that micro- and nanoplastics are harmful to human health, causing inflammation, oxidative stress, and immune responses. Emphasizing how larger polystyrene particles have been shown to cause inflammatory effects in lung cells, whereas unaltered or carboxylated polystyrene nanoparticles can up-regulate inflammatory genes in various cell types (Yee et al. 2021).

Acknowledging this, our modern lives have been heavily impacted by MP pollution as these particles have merged into many places that negatively affect our everyday lives. For example, MPs have been found in oceans, rivers, protected areas, food, drinking water, and almost every conceivable environment. Furthermore, several types of MP particles are usually found in wastewater treatment plant facilities as they typically originate from materials that are designed and largely utilized for commercial and household use, including textiles, detergents, pharmaceuticals, cosmetics, packaging, etc. These particles are generally invisible to the human eye, which is why typically special equipment like microscopes and Fourier-transformed infrared spectroscopy are used as methods to identify and characterize them (Karami et al. 2022).

As indicated, water and wastewater treatment plants are major pathways for MP particle's introduction into the environment. However, the understanding of the efficiency of these plants related to the retention and eventual elimination of MP pollution is an area of critical need for the potential improvement of decontamination methods (Cristaldi et al. 2020). In present study, that is part of a Distinction in the Major research certificate (Abumariam, 2023), a combination of literature review and research focused on exploring methods that are used for the characterization and quantification of MPs from wastewater treatment plants is employed and reported. Furthermore, we describe other aspects of this issue including, making a comparison between the quantity of MP found in each respective study in addition to the types of the different treatment methods previously implemented. In order to investigate these aspects, in the present study, samples from different units of Cookeville's Wastewater Treatment Plant have been taken to be analyzed so that the presence of MP particles can be studied. Experimental procedures were followed based on previous studies (Karami et al. 2022) to isolate possible MP particles from the samples taken and characterize them using both microscope and FTIR techniques for increasing our qualitative knowledge. This research was complemented by a quantitative analysis that contributes to the understanding of how the different units of the Cookeville, TN, WWTP plant is performing with respect to the retention of the MP before discharging the treated effluent into a local creek. Finally, the results of aforementioned experiments are discussed as well as other possible new techniques that could be implemented to enhance future studies' efficiency will be presented.

Refences

1. Turan, N. B., Erkan, H. S., & Engin, G. O. (2021). Microplastics in wastewater treatment plants: Occurrence, fate and identification. *Process Safety and Environmental Protection*, *146*, 77-84.
2. Yee, M. S. L., Hii, L. W., Looi, C. K., Lim, W. M., Wong, S. F., Kok, Y. Y., ... & Leong, C. O. (2021). Impact of microplastics and nanoplastics on human health. *Nanomaterials*, *11*(2), 496.
3. Karami, S., Rashidi, R., Arce, Pedro, A.E., Saraei, N., "Understanding the Transport and Fate of the Microplastic Particles in a Main Wastewater Treatment Plant: Case Study of a Typical Plant in Western Iran".
4. Karami, S., Rashidi, R., Arce, Pedro, A.E., Saraei, N., "Understanding the Transport and Fate of the Microplastic Particles in a Main Wastewater Treatment Plant: Case Study of a Typical Plant in Western Iran".
5. Cristaldi, Antonio et al. "Efficiency of Wastewater Treatment Plants (WWTPs) for Microplastic Removal: A Systematic Review." *International journal of environmental research and public health* 17,21 8014. 30 Oct. 2020, doi:10.3390/ijerph17218014
6. Kiran, Boda Ravi et al. "Micro/nano-plastics occurrence, identification, risk analysis and mitigation: challenges and perspectives." Re/views in environmental science and bio/technology vol. 21,1 (2022): 169-203. doi:10.1007/s11157-021-09609-6

173t Influence of Polymer Architecture on Catalytic Deconstruction of Polyethylene

Alex Balzer, Zachary Hinton, LaShanda Korley, Thomas H. Epps, III

Catalytic deconstruction has emerged as a promising solution to valorize polyethylene (PE) waste into valuable products, such as oils, fuels, surfactants, and lubricants. Unfortunately, commercialization has been hampered by inadequate optimization of PE deconstruction due to an inability to either truly characterize the solid reaction products or adjust catalytic conditions to match the ever-evolving product distribution (and associated property changes) during reaction. To address these challenges, a detailed analysis of molecular weight (MW) distributions and thermal characterization was developed and applied to PE deconstruction, which enabled more thorough identification of polymer chains within the solids (*e.g.*, changes in MW or branching structure). Critical distinctions between solids obtained from multiple hydroconversion reactions were captured. For example, solids generated from PE hydrocracking exhibited a broadened MW distribution with a disappearance of a significant fraction of highly-linear segments of the polymer chains. Additionally, the polymer feedstock architecture influenced the interactions with the catalyst and reaction kinetics. By tracking the deconstruction behavior as a function of reaction type, time, temperature, and catalyst design, we mapped critical pathways toward PE valorization. These results will enable the incorporation of changes in intrinsic polymer properties (*e.g.*, MW distributions, branching) into process-specific catalyst, reaction, and reactor-design calculations to enhance the economic potential of PE deconstruction.

173u Simultaneous PFAS Mineralization and Sorbent Regeneration Using PFAS Mineralizing Molten Salts (PMMS)

Emily Gonzales

Per- and polyfluoroalkyl substances (PFASs) are a group of synthetic organofluoride compounds that have generated public concern due to their adverse health effects and widespread persistence in the environment. Currently, PFAS-impacted wastewater streams are treated using conventional methods including filtration with granular activated carbon (GAC) and ion-exchange resins (IXR). While state-of-the-art physical adsorption methods are consistent and reliable for PFAS removal from water, these methods only transfer the PFAS from the water phase to the adsorbent material. This does not eliminate the contaminant and further results in secondary pollution. There is no existing widely implemented commercial technology for treating and regenerating these spent adsorbent materials. PSI's solution is to create a scalable, safe and economical method to treat PFAS-impacted adsorbent material that simultaneously mineralizes PFAS compounds and regenerates the adsorbent material under solvent free conditions using a novel PFAS Mineralizing Molten Salt (PMMS) platform. PMMS achieves high defluorination efficiencies by using a low-temperature molten salt reactor (MSR). The key innovation of the PMMS system is the application of a cost-effective molten salt eutectic mixture that has a low melting point (170 °C). The PMMS system concept allows for safe operation at ambient pressure, is readily scalable, and has very low treatment costs due to the inexpensive nature of the reagents and use of commercial off the shelf processing equipment. This poster describes the feasibility and preliminary experimental results from the PMMS system.

173v Catalytic Pyrolysis of Face Masks to Liquid Fuels

Zachary Shin, Eric Dudley, Amir-Hadi Boroumand, Adrian Rubio, Shiaojung Louie, Pilar Cuadros-Arias, Jacob Dermovsesian, Mingheng Li

Throughout the COVID-19 pandemic, 15 trillion face masks were used and discarded as plastic waste. To mitigate mask pollution, a catalytic pyrolysis method was employed to decompose face masks into hydrocarbons in the range of gasoline and diesel. A semi-batch reactor was designed and built to crack the face masks at various temperatures with and without the presence of the zeolite catalyst, yielding a crude oil product after condensation. The compositions of the products were then analyzed using two-dimensional gas chromatography. The results displayed a higher n-paraffin count for uncatalyzed products as compared to a higher iso-paraffins count in catalyzed product.

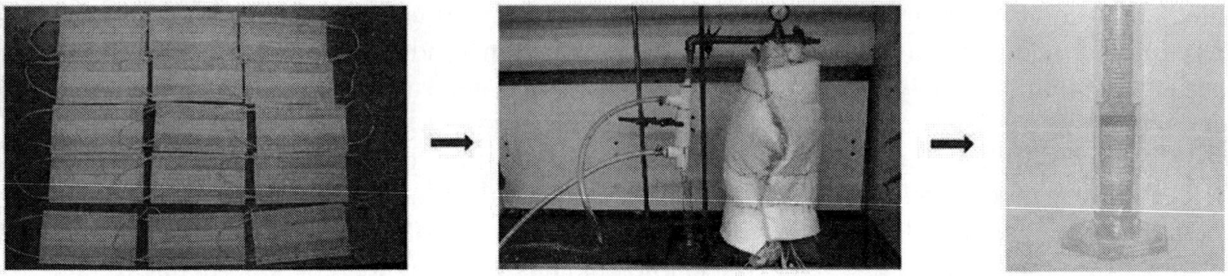

face masks | catalytic reactor | oil product

173w Optimization of the use of the River-Retention Pond System as an adaptive response to the Climate Change impact on water supply in rural Districts in the Caribbean

Adrian Lutchman, Festus Addo-Yobo

Optimization of the use of the River-Retention Pond System as an adaptive response to the Climate Change impact on water supply in rural Districts in the Caribbean.

Adrian Lutchman, F. Addo-Yobo*,

The Department of Process Engineering, The University of Trinidad and Tobago, Point LIsas, Couva, Trinidad and Tobago.

*** Corresponding Author**

Abstract.

Many Eastern Caribbean Countries suffer variations in rainfall patterns, with attendant periods of drought and excessive downpour and flooding partly due to the adverse impacts of climate change on local weather patterns. As a result it has become necessary to modify or create additional infrastructure to achieve resilient supply to meet the needs of the population. However, the conundrum here: is that the longer term impact of climate change is unpredictable therefore it is inadvisable to invest heavily in infrastructure which might become redundant or even prejudice future adaptive responses. It is for this reason that the OECD advocates the exploitation of existing resources or natural pre-existing facilities and adapting them to meet the requirements (New Zealand, OECD report).

In Caribbean as in Canada, a promising adaptive response is to harvest surface water in rivers and treat it in modular water treatment plants located close the districts for which there is chronic periodic water shortage. These areas also suffer inundation and flooding in many countries in the Eastern Caribbean area. Our previous work has used the PRMS (Precipitation Runoff Modelling system) within GSFLOW software to model the hydrology of watersheds and asses and predict water availability in the rivers (Lutchman, Addo-Yobo 2024) based on historic and future climate data, and existing surface cover of the land. The strategy has been based on using the river-retention pond system to provide water to a potable treatment plant as well as attenuating the risk of flooding downstream. This is less expensive for augmenting existing infrastructure to improve water supply resilience compared to building dams or artificial ponds. The present work uses chemical engineering techniques to determine the pond filling schedule that minimizes the risk of flooding integrated over time, subject to the constraint of ensuring the plant has adequate water for potable treatment.

This presentation exploits GSFLOW, an open source software, for mechanistic modelling of the hydrology of the Vega Oropouche watershed as a tool for managing the provision of surface water for potable treatment and reducing the risk of flooding based of a simplified objective function subject to various constraints. The Vega-Oropuche watershed has only one major river, an aquifer and several ponds which have been created by sand mining decades ago which can now be opportunistically harnessed for storage of water from the river for the above purpose and which in future can also be used for water retention to mitigate flooding.

Methods

The model requires time series input of precipitation and temperature, which are influenced by climate change. Global Circulation Models (GCM) and Regional Circulation Models (RCM) have been calibrated alongside weather generating algorithms to predict historic data and future data over a 30 year horizon (2024-2055) (Bachoo, Lutchman and Addo-Yobo). Whereas prediction of the impact of climate data on hydrology is essentially stochastic, this work employs deterministic methods in predicting the rainfall-runoff into the rivers. The variance of the error distributions has been approximated by linearized precipitation-runoff approximations (Farmer, William H., and Richard M., Vogel, (2016)).

Results

Table 1: Bathymetric and Lidar Studies of the Ponds

Sites	Surface Area (m2)	Average Depth(m)	Volume(Imperial Mgal)
Vega de Oropouche Pond	100,333	7	183
Tapana Pond	99,200	4.5	99
New Pond 1	68,500	WIP	WIP
New Pond 2	132,000	WIP	WIP
New Pond 3	120,110	4	105.7

Figure 1 shows the characteristic curves, depicting the evolution of the volume of water retained at the base case Pond, for the different filling regimes and, the capacities of the pristine and the dredged pond (shown respectively as solid and broken horizontal lines). The rates at which water has to be excluded from the pond, when it is full by shutting the valves to the pond at the weir on specific days are shown in red. The blue curve demonstrates that the dredged retention pond would meet the requirement of the pond to provide 4 million gallons per day of water for treatment and reserve space for excess flood water from the River to be held. For a few-days between 580th and 680th days and also on days 1074, 1357 and 1374, the valve at the weir would have to be closed to prevent the pond from filling beyond capacity. This represents less than 15% of water which could have been diverted to the pond, were it larger. The policy is such that 25% of the river flow is diverted to the pond area at low flow rates and 80% percent diverted at high flow rates.

Figure. 1: Evolution of volume of water retained at the RSSP Pond over a three years and ten-months.

The impact of the retention pond in mitigating the flooding of the river is demonstrated in Figure 2, which compares the flowrate of the river before and after the incorporation of the pond in the river system. The peak in the river flow around day 290 is significantly reduced, almost fivefold, from about 960 million gallons per day to 200 million gallons per day after the introduction, due to the action of the pond. There are similar reductions on days 291 and 292. Conversely, flowrates after the incorporation of the pond are higher than the flowrates before the incorporation of the pond around days 656 and 683.

Figure 2: Impact of the Base Case Pond on the flowrate of the River before and after diversion to the Pond.

Conclusion and Future work

Our work has shown that modelling of the characteristics of watersheds, the recharge from neighbouring rivers and ponds in locality of the Vega-Oropouche River in Trinidad and Tobago, is useful for assessing the feasibility of water projects, and their resilience to climate change.

The work has demonstrated conclusively that the inclusion of one retention pond of capacity 2.4 billion gallons (the base case) significantly impacts the daily flood risk and the integrated flood risk over a period of three years, when compared to the situation when the pond is not used. Future iteration will consider the deployment of the other ponds of with varying capacities in the River-Retention Pond scheme.

References

Farmer, William H., and Richard M., Vogel, (2016), "On the deterministic and stochastic use of hydrologic models", Water Resources. Res., 52, 5619–5633, 2016, doi:10.1002/2016WR019129.

UNEP (2000), "Dams and Development: A New Framework for Decision-Making". *The Report of the World Commission on Dams*, Earthscan, London

World Economic Forum, http://www3.weforum.org/docs/WEF_Global_Risks_2015_Report15.pdf

Adrian Lutchman, Festus Addo-Yobo, Amarnath Chinchamee, Shalini Ramsaran Baksh, and Daniel Sammy "Hydrological Modelling of Ravine Sable Watershed for Evaluation of Surface Water Supply for Treatment and Flood Mitigation

Adrian Lutchman and Festus Addo-Yobo, "Sourcing surface water from the Vega-Oropouche watershed for potable water treatment; an adaptive response to climate change in an eastern Caribbean Country", International Conference on Water Management and Modelling, February 2024, Ontario, Canada.

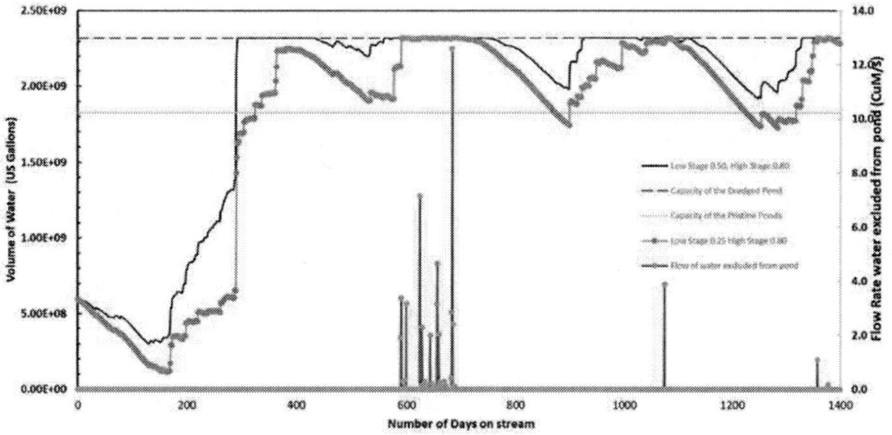

Figure. 1: Evolution of volume of water retained at the RSSP Pond over a three years and ten-months.

The impact of the retention pond in mitigating the flooding of the river is demonstrated in Figure 2, which compares the flowrate of the river before and after the incorporation of the pond in the river system. The peak in the river flow around day 290 is significantly reduced, almost fivefold, from about 960 million gallons per day to 200 million gallons per day after the introduction, due to the action of the pond. There are similar reductions on days 291 and 292. Conversely, flowrates after the incorporation of the pond are higher than the flowrates before the incorporation of the pond around days 656 and 683.

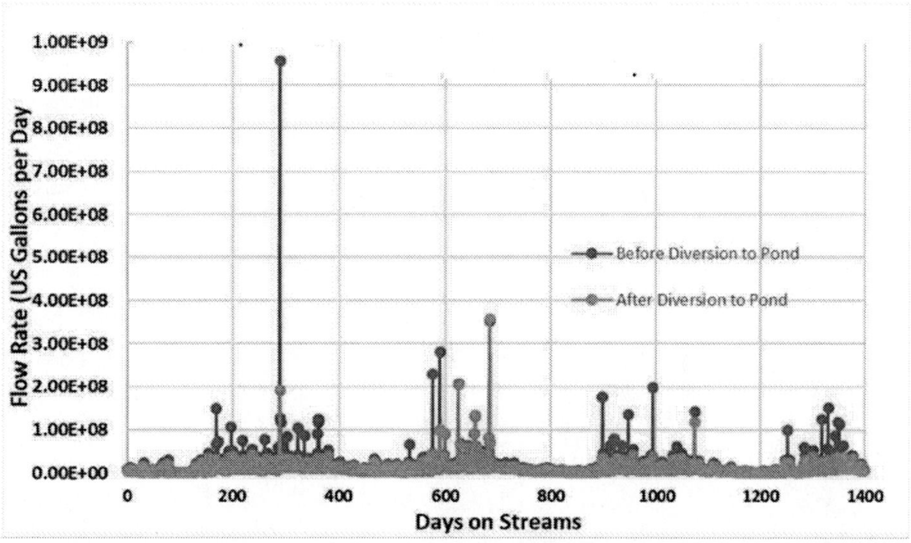

Figure 2: Impact of the Base Case Pond on the flowrate of the River before and after diversion to the Pond.

173x A Holistic Strategy for Treatment of Raw Landfill Leachate, Textile Contaminants and Pharmaceutical Wastewaters: An Electro-Peroxone-Based Approach

Ramya Srinivasan

Globalization and industrialization have led to immense environmental pollution. Solid waste generated by cities are of the order of billions of tons per year resulting in enormous landfill leachate generation. Untreated leachate is capable of polluting the neighboring ground water sources. On the other hand, presence of textile and pharmaceutical contaminants in water has increased rampantly, causing endocrine disruption and chronic toxicity to human and aquatic life. Conventional treatment technologies are unable to completely remove these pollutants resulting in contaminated water bodies.

Electro-peroxone, a propitious advanced oxidation process, is herein investigated in the treatment of a broad-spectrum of wastewaters, namely, raw landfill leachate, textile contaminants and pharmaceutical wastewater. In this study, the electro-peroxone system comprises of a Reticulated Vitreous Carbon (RVC) cathode and Platinum-coated Titanium anode. To understand the system better, quantification of *in situ* generated H_2O_2 (~150 mg/L) and $^{\cdot}OH$ radical was performed. This robust system resulted in the simultaneous removal of (i) Total Organic Carbon (TOC), (~70-90% based on nature of wastewater) (ii) Color (~99.7%), (iii) Disinfection (~99.9%), (iv) Organic suspended solids (~85%), (v) Turbidity (~80%) and hence (vi) Toxicity. The mechanism behind the efficient removal of various wastewater parameters was also investigated. Overall, a holistic treatment of various industrial contaminants and wastewaters is achieved with a removal efficiency ranging from 70% to complete removal, based on the nature and complexity of the contaminants present.

With the advantages of (i) retrofitting to an existing ozonation-based plant, (ii) sludge-free and (iii) chemical-free treatment, electro-peroxone proves to be a promising technology for the effective removal of such persistent contaminants present in complex real wastewater systems.

References:

Srinivasan, R., & Nambi, I. M. (2022). An electro-peroxone-based multi-pronged strategy for the treatment of ibuprofen and an emerging pharmaceutical wastewater using a novel graphene-coated nickel foam electrode. *Chemical Engineering Journal, 450*, 137618.

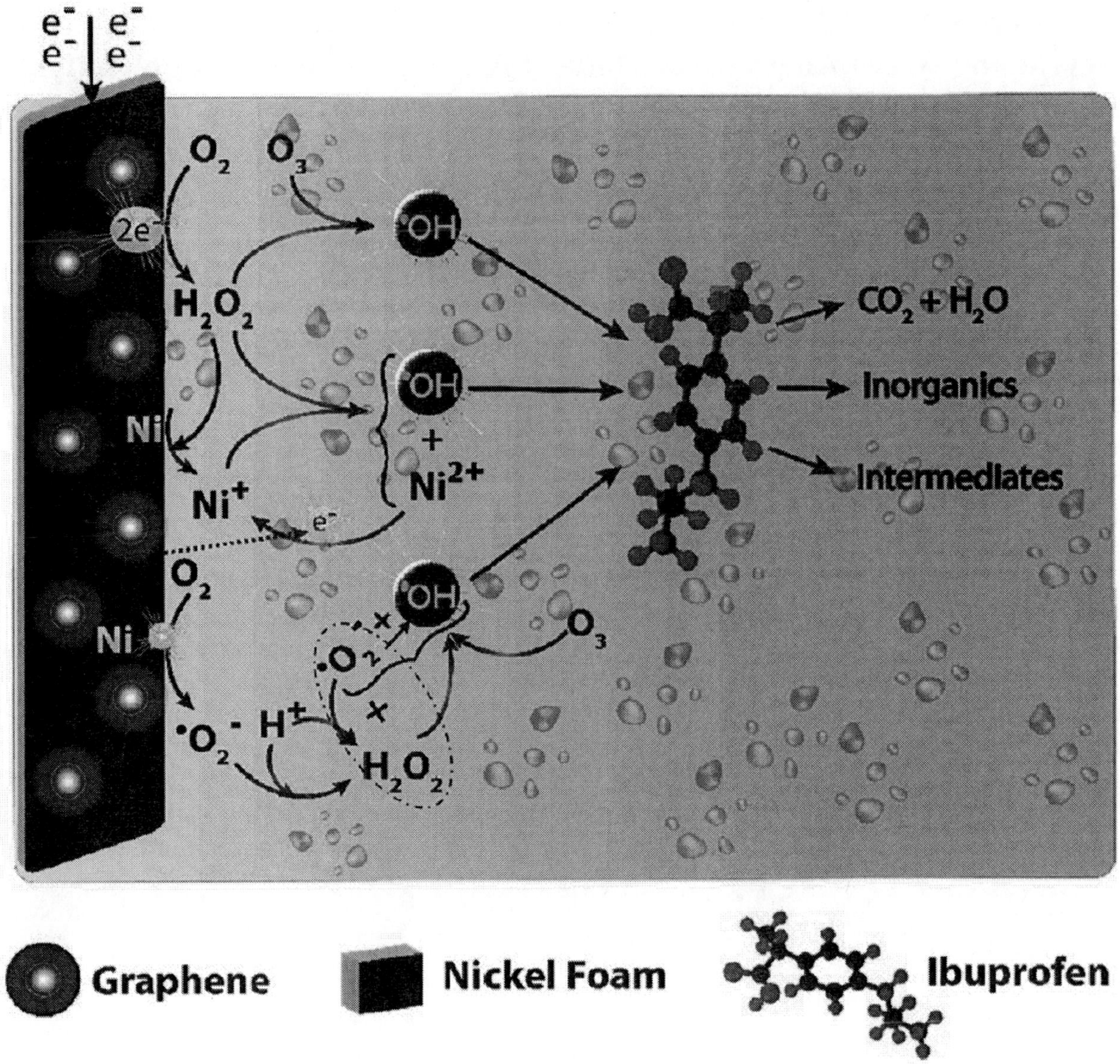
e⁻ e⁻
e⁻ e⁻
O2
O3
2e⁻
OH
H2O2
OH
+
Ni2+
Ni
Ni+
e⁻
O2
OH
Ni
O2
O3
O2
H+
H2O2
CO2 + H2O
Inorganics
Intermediates
Graphene
Nickel Foam
Ibuprofen

173y A Computational Study on Groasis Waterboxx Technology: The Feasibility of Developing Green Alternatives Made of Poly-Lactic Acid and Polybutylene Succinate

Heriberto Garcia, Mohammad Davachi, Miaad Dastjerdi, Arman Mohammad-Ali Dastjerdi, Pegah Mohammad-Ali Dastjerdi, Javad Dastjerdi, Behzad Shiroud Heidari

Developing water-saving technologies like the Groasis Waterboxx (GW) using environmentally friendly materials presents a reasonable idea to prevent the harmful impacts of the agricultural industry on the environment and society. In this study, we simulated the injection molding process of green GWs made from poly-lactic acid (PLA) and polybutylene succinate (PBS) to analyze their performance under rigorous conditions. Using Autodesk Moldflow, we investigated how different process parameters affected cycle time, energy consumption, and the final product's mechanical and visual qualities, comparing them with polypropylene (PP). Additionally, we simulated GW performance in the Sahara Desert using Abaqus. The simulations showed that PLA had shorter injection and cooling times compared to PP, resulting in a 44% and 56% reduction, respectively, indicating faster production. While PBS-based parts exhibited mechanical qualities similar to PP, PLA demonstrated lower shrinkage and warpage. However, environmental simulations suggested that PLA might not be practical in areas with high thermal stresses due to its glass transition temperature overlapping with ground temperature, leading to potential failure near the land surface. Despite its biodegradability, PBS demonstrated a high safety factor of approximately 82% compared to PP samples. Although the injection of PBS consumes more energy due to higher required pressure, replacing PP with PBS is justified considering its efficiency and eco-friendliness. These findings align with experimental results reported by the manufacturer and emphasize the importance of considering both performance and environmental factors when developing sustainable agricultural technologies.

173z Bromate-Free Disinfection of Bottled Water Using High Voltage Impulse

InSoung Chang

Disinfection using ozonation for bottled water usually produces various kinds of disinfection byproducts (DBPs). Among the DBPs, bromate (BrO_3^-) is notorious for its potential carcinogenic nature. World Health Organization (WHO) limits the bromate to 10 ppb in drinking water. To reduce this bromate formation during the disinfection processes, a high voltage impulse (HVI) technique was applied in this study. HVI has been used as one of the disinfection processes for non-thermal sterilization of foods. The HVI system consisted of a power supply, high voltage generator, capacitors, and impulse generator. HVI with 10 to 20 kV/cm of electric fields was introduced to sample waters to determine if bromates were formed. As the precursor of bromate, i.e., bromide (Br^-) spiked into the sample waters by 1,000 ppb, bromate was not generated at all times, indicating that the bromate formation was completely inhibited. A model microorganism, *Escherichia coli* is inoculated to the sample waters to test the disinfection efficiency. More than 2-log reduction (99 %) of *E.coli* was achieved after HVI treatment with 20kV/cm, 100Hz, 4μs pulse width. Consequently, the HVI technique could be used as an alternative strategy for bromate free-disinfection procedures for water disinfection.

173aa Molecular Dynamics Predicted Selenium Extraction to Reduce Selenate Toxicity in Wastewater

Tridip Das, Rohit Srikanth, Paresh Samantaray, William A. Goddard III

Selenium (Se) is an essential micronutrient for animals and humans at low concentrations, but it becomes extremely toxic at higher levels. Its presence in the environment arises from both natural sources (such as volatilization of water bodies, terrestrial weathering, and volcanic activities) and anthropogenic activities (including fossil fuels, mining, and oil refining). In water systems, selenate (SeO_4^{2-}) and selenite (SeO_3^{2-}) are the predominant Se species. Selenate, being highly soluble and poorly adsorbed by sediments and soil, poses a significant challenge due to its mobility and potential bioaccumulation. Selenite, though less soluble, can also accumulate in aquatic ecosystems. The consequences of selenium toxicity are far-reaching. Acute exposure to Se compounds can lead to neurological issues, respiratory problems, and other health complications. Chronic exposure has been linked to cancer development, decreased sperm movement, and disruptions in insulin signaling. Therefore, understanding the occurrence and effective management of selenate and selenite in water is critical for safeguarding human health and the environment.

In our computational study predicting selenate breakdown to selenium at different pH using AIMD involves initializing the system with selenate ions, introducing excess electrons, and allowing the dynamics to evolve. By analyzing trajectories and electronic properties, we have gained the insights into the reduction process of selenite/selenate and the formation of selenium species in steps like Se2 à Se4 à Se8. We have adjusted H+ concentration and convergence criteria appropriately to accurately predict reaction conditions to precipitate Se8 from water. Based on the simulation guided pH, experiments were performed and we were able to precipitate Se8 from selenate containing water at room temperature.

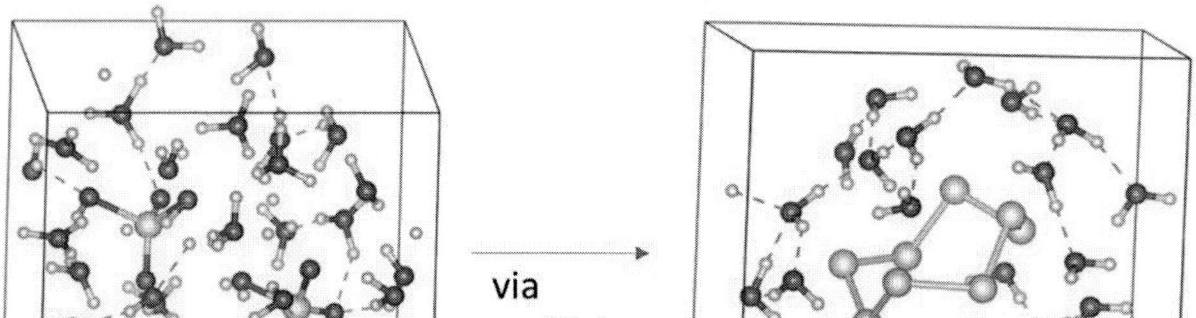

Green – Se, Red – O, White – H

173ab A Study in Mexico for Advancing the Circular Economy and Sustainable Waste Management through Optimization of Waste-to-Energy Technologies

Ilse María Hernández-Romero, Lucy T. González, Antonio Flores-Tlacuahuac

This study introduces an approach to sustainable waste management through Non-Linear Programming (NLP) for multi-objective and multi-period optimization, effectively integrating waste-to-energy technologies. The study examines the potential of Municipal Solid Waste (MSW) management as a significant energy reservoir by utilizing landfill gas recovery, biofuel production, and energy production. To comprehensively assess waste-to-energy options, the model incorporates four distinct technologies: anaerobic digestion, incineration, gasification, and pyrolysis. It carefully considers thermochemical, technical, economic, environmental, and social factors using rigorous non-linear functions, thereby offering a balanced strategy for waste management. Focusing on Mexico City, the research demonstrates the ability of the model to align economic goals with environmental sustainability. The methodology stands out for its capacity to balance the dual objectives of reducing environmental impact while enhancing economic benefits. By utilizing technologies like gasification and incineration, the model not only highlights the potential for significant emission reductions but also underscores the importance of energy production, contributing to sustainable energy goals. This work makes a significant contribution by providing a scalable and adaptable framework for sustainable waste management decision-making, advancing the field, and supporting urban sustainability initiatives.

173ac Preparation of Carbon Electrodes from Food Waste for Sodium-Ion Batteries

Isamu Umeda, Rajesh Shende, Sandeep Kumar

Sodium-ion batteries (SIBs) are promising as the next generation of batteries. Sodium is a low cost, abundantly available element and has shown rapid rechargeability. Unlike lithium-ion batteries, which uses graphite as an anode material, SIBs can use hard carbon. Hard carbon is the amorphous carbon that cannot be graphitized even at more than 3000°C, can capture and store more sodium ions. Therefore, hard carbon that can perform well as SIB anode was desirable. More importantly, it should be preferable if it can be produced from sustainable materials. In this study, we aim to produce food waste derived hard carbon via heavy bio-oil with supercritical organic solvent treatment and analyze the fundamental characteristics of the produced hard carbon. The food wastes were converted into heavy bio-oils by heating with organic solvent such as methanol, ethanol, and acetone under supercritical condition. The bio-oils collected were solidified into carbonaceous material through vacuum heating. Finally, the carbonaceous materials were produced through activation process. The hard carbons were characterized by analyses such as proximate and ultimate analysis, FTIR, BET, TGA, ICP-OES, SEM, and cyclic voltammetry. The effects of different types of organic solvent on their structures and electrochemical performances were discussed based on the analytical results.

173ad Effective Anodic Sulfide Removal Catalyzed By Nickel Phosphide Modified Stainless Steel Electrodes

Xin Zhang, Riying Qi, Kuichuan Sheng, Hongjian Lin

Hydrogen sulfide (H_2S) is toxic and widely exists in sulfide-containing waste and wastewater treatment processes, such as septic tanks, sewer systems for municipal wastewater, livestock manure storage pits, and anaerobic digestion, etc., causing odor issues and life-threatening problems. The concentration of H_2S in biogas varies from 50 ppm to 5000 ppm, while the Immediately Dangerous to Life or Health (IDLH) is 100 ppm. Removing H_2S to reduce work-related injury and environmental corrosiveness due to H_2S presence is critical work. Several methods, e.g., salt precipitation, adsorption, bio-scrubbing, and anode sacrifice, were developed for sulfide control in different scenarios. High sulfide removal efficiency has been achieved by these methods, but these methods tend to consume a large amount of chemical and energy. Alternatively, electrochemical sulfide oxidation is a promising approach for sulfide removal from wastewater or biogas due to its high sulfide removal efficiency, low energy, and chemical consumption. In this work, nickel phosphide was facilely electrodeposited on stainless steel (SS/Ni-P) and comprehensively evaluated as an anode catalyst for sulfide oxidation. Electrochemical property analysis suggested that SS/Ni-P had low polarization resistance (109.10 Ω) and high electrochemical active surface area (C_{dl} = 1.3 mF/cm^2). The sulfide removal efficiency of SS/Ni-P achieved 95.00 ± 3.55 % at 1.42 V *vs* RHE within 48 h, which was 2.60 times and 3.62 times of that by SS/Ni and SS anodes, respectively. The first-order kinetic rate constant of SS/Ni-P for sulfide removal was 0.0315 ± 0.0061 h^{-1}, superior to the reported nickel-based catalyst with stainless steel substrate. Pre-treating SS/Ni-P under a constant external potential for 10 s further improved the electroactivity on sulfide oxidation due to the enhancement of the electrochemical active surface area by 46.56%. Further analysis of the chemical state changes during the sulfide oxidation process revealed that the Ni from Ni-P was the real active center, which formed Ni-S precursors to promote polysulfide formation. This work provided a highly efficient sulfide oxidation anode material for sulfide-containing wastewater and biogas treatment.

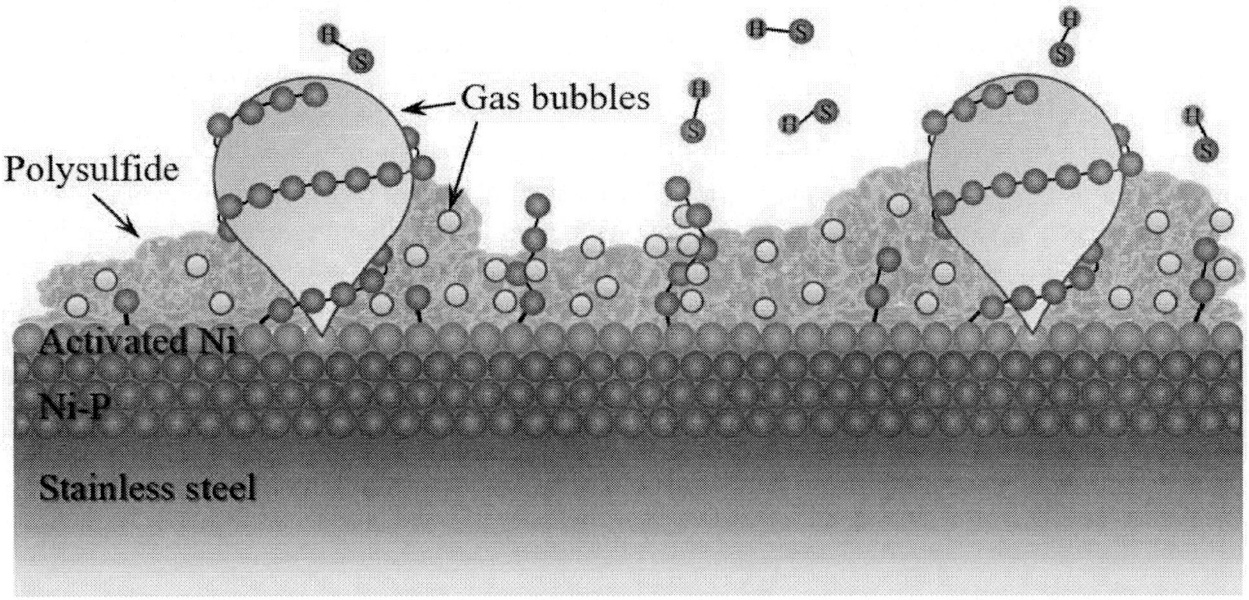

173ae Tracking and Predicting Regional Greenhouse Gas Emissions: A Case Study of Connecticut

Peiyao Zhao, Jimi Oke

Global warming triggers wildfires, rising sea levels, and habitat loss for many species. This is driven by elevated greenhouse gases concentrations in the atmosphere. Emissions tracking and prediction frameworks are important to understand the source contributions and find mitigation strategies. Compared with developing a nationwide emissions inventory, however, a regional-level emissions inventory database is crucial for local governments to identify the specific activities and processes responsible for the emissions and implement tailored policies to reduce GHG emissions. This study develops a GHG inventory estimation and projection platform tailored specifically for the local area. This accounting method was applied to three metropolitan statistical areas (MSAs) in Connecticut, with the aim of discovering the emissions characteristics at different locations and forecasting future emissions shift and trends. However, the challenge of developing an inventory at this regional level lies in the availability of open and granular data. To address this issue, we propose a hybrid accounting approach, utilizing a bottom-up estimation based on activity data, which is supplemented by a top-down method by scaling up statewide emissions. We found that emissions in Hartford, Bridgeport, and New Haven MSA totaled 10.95, 7.4, and 6.97 million metric tons of carbon dioxide ($MMTCO_2e$) in 2021, respectively. Among all the emissions from seven sectors (transportation, electricity consumption, solid waste, stationary combustion, agriculture and wastewater), fossil fuel related activities, such as on-road transportation and residential heating, were the primary source of emissions. In each MSA, the transportation sector contributed to the highest amount of emissions, followed by stationary combustion. Besides, the percentage of electricity consumption in each MSA exceeded that of the entire state, illustrating the higher electricity consumption in urban area, which are heavily centralized in those three MSAs. The per capita emissions were 10.95, 7.41, and 9.6 $MTCO_2e$ in Hartford, Bridgeport, and New Haven, respectively while the average per capita emissions for the entire state was about 9.59 metric tons. Surprisingly, Hartford's per capita emissions exceeded that of the entire state, suggesting the need for specific emissions reduction measures in this area. Additionally, the results show that trees play a significant role in sequestering carbon from the air, with 0.62, 1.98, and 0.74 $MMTCO_2e$ in the three areas, respectively. This illustrates the need to maintain our forestry and promote greenery. After the establishment of the inventory processing and calculation procedure, historical data and emissions factors were utilized to estimate past emissions. We applied the auto-regressive integrated moving average (ARIMA) model to predict the future GHG emissions across all sectors up to 2035. Comprising autoregressive, integrated, and moving average terms, this model forecasts future values based on historical data. The model parameters (p,d,q) were determined through parameter estimation and model testing to minimize the root mean squared error. The resultant prediction outcomes offer valuable insights for policymakers to formulate strategies and set emission reduction targets. We expect that our framework would serve as a valuable tool for regional emissions accounting in support of nationwide net-zero goals.

173af Study of Potential Environmental Impacts during Traffic Accidents Involving Vehicles Transporting Liquid Hydrocarbons

Natalia Godoy Silva, Alvaro Orjuela, David E. Herrick

In Colombia, transportation of hazardous chemicals is frequently carried out using trucks that commute throughout the country. In particular, liquid hydrocarbons such as crude oil, gasoline and diesel stand out due for their high frequency of transportation and the large volumes. The land transport of these substances entails an inherent danger and risk for drivers and for the communities and ecosystem surrounding the road corridors. In order to evaluate the possible risks and environmental impacts that an incident involving the spill of dangerous chemical substances would entail (particularly for liquid hydrocarbons), a computer-based assessment was carried out using open software tools. Initially, an analysis was done to identify the hydrocarbons that represent the greatest risk, as well as the road corridors with the highest accident rates. This was done by using a potential risk estimation using a novel methodology involving quantitative assessment based upon the Globally Harmonized System of Classification and Labelling of Chemicals (GHS) (Narváez et al., 2020). Subsequently, the most appropriate open software tool (ALOHA®) was selected to implement risk studies in the transportation sector (Jones et al., 2013). Despite its versatility, ALOHA® only allows modelling pure substances, and a methodology was developed to select suitable surrogate compounds to described the studied hydrocarbon mixtures. Once selected, a scenario analysis was carried out to model an incident involving the release of fuel in a high-accident rate corridor in the country. The effect of using surrogate substances and substances with user-defined properties was studied. Liquid release, vapor cloud dispersion, and fire events were also evaluated. This made it possible to evaluate affected areas, risks of chemical exposure and radiation damage, as well as to estimate midpoint environmental impacts based upon reported emission factors (Bare, 2012).

Based upon historical records of transported hazardous materials, it was found that 95% corresponded to hydrocarbons including crude oil, fuels, solvents and petrochemical feedstocks. Among these, around 30% corresponds to crude oil mainly transported from the east part of the country to the main refineries in north and central regions, whereas ~15% corresponded to gasoline. After application of a reported methodology for the preliminary evaluation of inherent hazard and potential risk, it was determined that gasoline and crude oil were also the substances that represent the greatest potential impact on the environment and the greatest risk in the social dimension. Thereby, crude oil and gasoline were used in the subsequent risk assessment.

Once the study case products were identified, the surrogate substances for the modelling in ALOHA® were selected. After minimization of the difference in the properties of interest (i.e. those used in spill, heavy-gas dispersion and pool fire models), it was found that local crude oil (i.e. Castilla crude oil) and gasoline could be adequately represented by using Cyclohexane, Ethylbenzene or n-Octane as surrogates. An alternative approach was also used by creating a new component in the data base with the properties of the studied hydrocarbons, but major issues occurred because ALOHA® do not allow to modify vapor pressure nor density of the materials. This generated a misprediction of spilled volume and of the evaporation rates during loss-of-containment events. This has been overlooked in previous studies and it is considered a major source of uncertainty when using ALOHA® for risk assessment. Finally, geographical information and data from a real truck accident was used to assess the goodness of the model. It was found that the software enabled to accurately predict some of impacts that occurred in the studied event. This indicate that the developed methodology and the use of ALOHA were suitable for conducting preliminary risk assessments and to develop contingency plans in the management of accidents involving transportation of liquid hydrocarbons.

References

Narváez, P. C., Serna, J., Orjuela, A., Camargo, M. (2020) Sustainability assessment for chemical product and process design during early design stages. In: Towards Sustainable Chemical Processes- Applications of Sustainability Assessment and Analysis, Design and Optimization, and Hybridization and

Modularization. P. 3-41. https://doi.org/10.1016/B978-0-12-818376-2.00001-6

Jones, R., W. Lehr, D. Simecek-Beatty, R. Michael Reynolds. 2013. ALOHA® (Areal Locations of Hazardous Atmospheres) 5.4.4: Technical Documentation. U. S. Dept. of Commerce, NOAA. Technical Memorandum NOS OR&R 43. Seattle, WA: Emergency Response Division, NOAA. 96.

Bare, J. C. 2012. Tool for the Reduction and Assessment of Chemical and Other Environmental Impacts (TRACI), Version 2.1 - User's Manual; EPA/600/R-12/554

Figure 1. Distribution of total number of transportations of liquid fuels in Colombia for 2018 and 2019

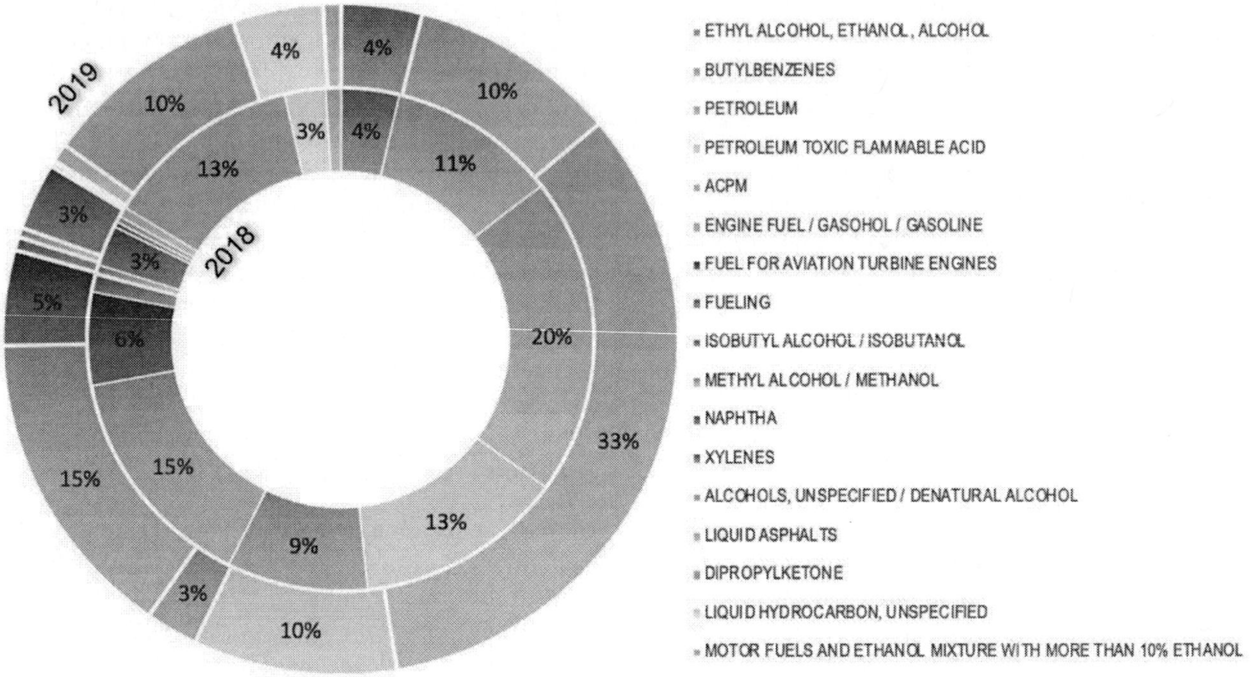

173ag Utilizing Mg-Bearing Minerals for Selective Nickel and Iron Recovery Coupled with Carbonates and Amorphous Silica Production

Ning Zhang, Ah-Hyung Alissa Park

This study addresses critical mineral scarcity and climate change by developing a sustainable approach for recovering multiple metals from olivine while promoting carbon sequestration. Through a comprehensive leaching process, metal recovery, and reagent recycling, the approach achieves selective recovery of valuable metals (Ni^{2+}) and converts Mg^{2+} into carbonates for CO_2 storage. Hydrometallurgical leaching facilitates over 90% metal dissolution and conversion, followed by Fe extraction and pH swing for high-purity products, such as iron hydroxide, amorphous silica, nickel carbonates and Mg-carbonates. It was found that iron removal can minimize metal loss for the nickel extraction and carbonation. Moreover, A techno-economic assessment highlights the economic benefits of this approach to removing iron in the early stage. Selective Ni^{2+} extraction is achieved using Versatic 10 by adjusting the equilibrium pH to 6.5-6.8 after removing impurities (e.g. Fe, Si), followed by stripping with bubbling CO_2. Mg-rich leachate is then used for Mg-carbonate production under ambient conditions. The proposed integrated process yields high-purity value-added products and sequestered CO_2.

173ah Assessing Global Warming Impacts on Food and Macronutrient Supply By Compartmental Modeling Approach

Sinue Arnulfo Tovar-Ortiz, Pablo Tenoch Rodriguez-Gonzalez, Rigoberto Tovar-Gomez

In recent decades, population growth, anthropogenic activity, and inefficient resource management have led to an excess of Greenhouse Gas (GHG) emissions and concentration in the atmosphere, intensifying the greenhouse effect and giving rise to phenomena such as Global Warming and subsequent Climate Change.In this context, the livestock sector, despite not being considered a direct emitter, significantly contributes GHG emissions with high global warming. Additionally, the agricultural sector is strongly influenced by livestock demand, as 80% of global agricultural production is destined for animal consumption. The increase in agricultural demand has exceeded sustainable production capacities, leading to practices such as deforestation and the exploitation of wild ecosystems to meet this growing livestock demand.

In addition to its environmental impacts, the intensive focus on meeting livestock demand has significant repercussions for macronutrient supply and distribution. The emphasis on animal consumption within the agricultural sector not only diverts resources away from direct human consumption but also affects the availability and diversity of macronutrients in the global food supply. This skewed allocation of resources towards animal feed production limits the accessibility of essential macronutrients such as carbohydrates, proteins, and fats for human consumption, particularly in regions where food insecurity is prevalent. Moreover, the reliance on monoculture practices to sustain this demand further exacerbates the issue by compromising the nutritional diversity of crops grown for both human and animal consumption. Thus, while addressing the environmental implications of livestock production is crucial, it is equally imperative to consider the broader implications for macronutrient availability and nutritional equity on a global scale.

Considering these complex interrelations between anthropogenic activities, environmental impacts, and macronutrient supply, the utilization of mathematical models, particularly compartmental models, emerges as a valuable tool for analysis and prediction. Compartmental models allow researchers to represent the intricate dynamics of various sectors and their interactions within a unified framework, enabling a comprehensive assessment of the repercussions of livestock demand on macronutrient availability and environmental sustainability. Furthermore, compartmental models facilitate the exploration of feedback loops and nonlinear relationships inherent in complex systems, offering insights into the long-term consequences of current trends and identifying critical leverage points for intervention. The objective of this work is to provide a tool that facilitates the analysis of strategies based on evidence-based decision-making and guides efforts toward achieving a more sustainable and equitable food system amid the challenges of climate change and growing global demand for food.

A mathematical global food model is proposed consisting of 9 compartments depicting the flow of matter between the primary economic sectors and adjacent sectors that contribute in some way to the human food chain. The agricultural sector is divided into two compartments: Grains, which are edible for humans, and Forage, or residues that are not edible for humans. On the other hand, the Livestock sector is divided into Ruminant animals that generate a greater amount of GHG emissions and monogastric animals with a lower environmental impact. The interactions between compartments are modeled by Lotka-Volterra type differential equations and parameterized using the Food and Agriculture Database (FAOSTAT).

Additionally, the CO2eq concentration in the atmosphere is determined by summing the contributions of various greenhouse gases emitted or absorbed within the system. This calculation involves multiplying the compartment size by its respective Emission Factor to quantify the greenhouse gases emitted by each compartment. This GHG concentration feeds back into the model by influencing temperature changes and altering plant growth rates in the Lotka-Volterra equations. Finally, to calculate the food demand of the human sector, an average nutritional contribution is estimated for each unit of agricultural and livestock food. This estimation is obtained from FAOSTAT, utilizing the values of annual worldwide production and per capita consumption for each category in the year 2022.

The model is implemented in Matlab software, where parameters and differential equations are inputted and integrated using the Euler method with a unit step size over a span of 80 years. The results compares the per capita availability of macronutrients within the context of food availability. This comparison contrasts the requirements established by the WHO with the current macronutrient. The analysis reveals that presently, we are meeting the required parameters. However, projections indicate a shortfall by the year 2078, which would limit the availability of carbohydrates and minimum calories per capita.

In conclusion, the utilization of mathematical models, such as the one implemented in this work, offers invaluable insights into the dynamics of macronutrient supply, greenhouse gas emissions, and their implications for global food security and environmental sustainability. However, the projections highlight the urgency of proactive intervention to address future shortages in macronutrient availability. Sustainable resource management practices, coupled with efforts to reduce greenhouse gas emissions and enhance agricultural productivity, are imperative to ensure a resilient and equitable food system for future generations.

173ai Co-Hydrothermal Carbonization of Biodegradable Straw Waste and Rice Husk

Thu-Huong Le, Juwon Kang, Hyungseok Nam, Doyeon Lee

Hydrothermal carbonization (HTC) is an environmentally friendly thermochemical process used to convert wet biomass into a solid carbon-rich product. Recently, there has been growing interest in co-hydrothermal carbonization (Co-HTC), which involves using two or more types of solid waste simultaneously. This approach is gaining attention as an efficient means of waste disposal. In this study, biodegradable straw waste (PLA) and rice husk (RH) were employed for co-hydrothermal carbonization. The investigation aimed to assess the impact of Co-HTC between biodegradable straw waste and rice husk at different ratios, reaction times, and temperatures on product distribution and hydrochar properties. The findings revealed several significant trends: 1) PLA/RH ratio: Increasing the RH content in the mixed feedstocks resulted in higher solid yields but lower oil yields. Additionally, the elevated RH content in the mixed feedstocks was proposed to promote the conversion of organic matter into hydrochar, leading to increased thermal stability and higher heating value (HHV) of hydrochar. 2) Reaction time: Prolonging the reaction time led to an increase in solid and gas yields, while the oil yield decreased. The longer reaction time enhanced cracking and recombination/repolymerization of molecules, resulting in higher volatile matter content in hydrochar and increased HHV. 3) Temperature: Higher reaction temperatures led to the maximum solid yield and HHV of hydrochar, achieved at 260°C. Overall, the study provides valuable insights into optimizing Co-HTC processes for efficient waste utilization and hydrochar production.

173aj An Analysis of Abandoned Mines in Texas

Emmanuel Dada, Mackenzie Stephney, Temitope Esan, Carol Akpan

Background: There are many mines located throughout Texas, totaling about 353 sites. Mines main purpose is to extract useful resources and minerals from the earth including copper, lead, uranium, and coal. Many mine sites had incomplete mining closure or improper reclamation that resulted in harmful effects on the soils, plants, crops, animals, and water. **Objective:** The objective of this study was to evaluate uranium mines and identify various remediation methods to abolish the environmental concerns in Texas. **Methods:** In this qualitative study, data was collected and analyzed using various resources. Twenty-three uranium mines were studied and examined in this research. **Results**: It was found that uranium mines contaminated its area through radioactive dust and toxins. Those issues have been swept under the rug due to Texas regulating flexibility for mining companies to set its own regulations. **Conclusion:** The prices and demand for uranium have fallen dramatically since 2007. Since then, many uranium mining sites have shut down and left unwanted hazards to the environment. To protect the general welfare and environment, government authorities and local communities are working to address these problems with different remediation programs.

173ak Phytoremediation of soils contaminated using Phaseolus vulgaris L and amendments

Alejandra Vargas-Beltrán, Julián E. López, Maria P. Reyes, Adriana Jaimes, Juan F. Saldarriaga

Anthropogenic activities, encompassing extensive agricultural and industrial operations across the globe, exert substantial pressure on the environment, culminating in profound ecological impacts. Particularly concerning is the escalating issue of soil pollution, an emerging threat resulting from the discharge of pollutants into terrestrial ecosystems. Among these pollutants, heavy metals like mercury (Hg) and agrochemicals such as chlorpyrifos (CPF) stand out due to their widespread presence and detrimental effects. Their accumulation in soils has now reached critical levels, jeopardizing soil fertility and posing significant consequences for both ecosystems and human well-being.

In response to this pressing challenge, the current study focuses on evaluating the effects of these contaminants on Phaseolus vulgaris L. (common bean). The study aims to explore the efficacy of enhancing the remediation capabilities of Phaseolus vulgaris L. through the integration of biochar, compost, and mycorrhizae amendments. This innovative approach to phytoremediation holds promise as a sustainable solution for mitigating soil contamination by Hg and CPF, offering a multifaceted strategy that addresses both organic and inorganic pollutants.

The soil was removed from Segovia, Colombia, a municipality known for its heavy mining activity (more than 9000 hectares), heavily contaminated soils from Hg use, and agricultural use. pH 5.85, CEC of 158 cmol(+)·kg-1, CRA of 50.04%, humidity of 4.50%, and organic matter content of 19.73% were the original soil's measurements. To determine the Hg content and confirm the amount of metal in the soil, three arbitrary samples were taken. The average Hg concentration of the entire soil was found to be 24.56 mg·kg-1. After the soil was added to a rizhobox, seedlings of Phaseolus vulgaris L. that had already germinated and measured on average 9 cm in height were planted. Throughout days 15, 30, and 45, chemical and physical properties like pH, electrical conductivity, and organic matter were observed, along with the bioavailability of Hg and CPF. The soil has been contaminated with different concentrations of chlorpyrifos (CPF), an organophosphate pesticide, at three specific levels: 40 mg·kg-1, 80 mg·kg-1, and 120 mg·kg-1. These values were selected to represent a range of contamination conditions that may be encountered in agricultural settings and other areas impacted by pesticide use. By employing these varied concentrations, the objective is to comprehensively evaluate the effects and efficacy of the proposed remediation methods, allowing for a detailed comparative analysis.

It has been found that the combination of biochar with mycorrhiza has the highest efficiency in CPF degradation, while biochar has the lowest bioavailability of Hg. Although the control treatment shows a degradation of CPF over time, the application of amendments favors a greater degradation of this pollutant. While in the case of Hg, the control also decreases the bioavailability of Hg over time, it is the application of amendments that considerably decreases the bioavailability of Hg. In the case of the application of mycorrhiza alone, it shows that the bioavailable concentration of Hg is higher than in the other amendments, and this is reflected in the concentration of the metal in the plant, which presents a greater amount of the contaminant in the root, showing similar results to those reported in the literature (Cruz et al., 2021; Leudo et al., 2020; Saldarriaga et al., 2023).

The study offered thorough insights into mycorrhizae and biochar's (B+M) remediation potential for CPF and Hg-contaminated soils. The results consistently showed that a combination of biochar and mycorrhizae was the most effective treatment for lowering the amounts of these contaminants. The degradation process was greatly accelerated by the synergistic effect of this combination. The B+M treatments demonstrated impressive reduction levels; at 80 mg/kg concentrations, they routinely achieved about 70% reduction, and at 120 mg/kg doses, they frequently approached 90% reduction. The important role these amendments play in the bioremediation process is highlighted by their great efficacy. Natural deterioration caused a slight drop in CPF concentrations in the control samples, but throughout the trial, the control group's CPF levels were always the highest when compared to the other treatment groups. This result emphasizes the benefits of employing mycorrhizae and biochar over natural degradation alone, since the latter is unable to achieve appreciable reductions in pollutant levels in a reasonable amount of time.

References

Cruz, Y., Villar, S., Gutiérrez, K., Montoya-Ruiz, C., Gallego, J.L., Delgado, M. del P., Saldarriaga, J.F., 2021. Gene expression and morphological responses of Lolium perenne L. exposed to cadmium (Cd2+) and mercury (Hg2+). Sci Rep 11, 11257. https://doi.org/10.1038/s41598-021-90826-y

Leudo, A.M., Cruz, Y., Montoya-Ruiz, C., Delgado, M. del P., Saldarriaga, J.F., 2020. Mercury Phytoremediation with Lolium perenne-Mycorrhizae in Contaminated Soils. Sustainability 12, 3795. https://doi.org/10.3390/su12093795

Saldarriaga, J.F., López, J.E., Díaz-García, L., Montoya-Ruiz, C., 2023. Changes in Lolium perenne L. rhizosphere microbiome during phytoremediation of Cd- and Hg-contaminated soils. Environmental Science and Pollution Research 30, 49498–49511. https://doi.org/10.1007/S11356-023-25501-Y/METRICS

173al Synthesis of biochar using economical pyrolysis processes as a remedial soil amendment

Juan F. Saldarriaga, Thomas Käslin, Julián López

Post-harvest residues are greatly undervalued across Latin American countries, often either burned or left to decompose in piles, which contribute to the generation of greenhouse gases. Therefore, it is essential for the health of soils, the livelihood of farmers, and the protection of the environment to seek low-cost and efficient solutions for better utilizing these substantial agricultural by-products. This paper evaluates two cost-effective technologies for producing biochar: Kon-Tiki and TLUD (Top-Lit UpDraft), both of which have demonstrated efficiency.

Following the collection and processing of a corn crop, the equipment was put on a farm in the Colombian municipality of Flandes. To achieve this, corn cobs were taken and placed in each device, causing a layer to build at the bottom before being ignited. It has been added capable of biomasses after being left to produce flame and calcines. As more layers of feedstock are added, the temperature gradually drops below the primary pyrolysis zone and, depending on how long the batch is left in the furnace, eventually reaches a final quenching temperature of 150°C to 450°C. There are two techniques for quenching. Depending on the furnace and how tightly the quench layer is sealed, the temperature of the biochar in the first process can stay over 400 °C for more than a day. By slowly pouring water from the top for the TLUD or injecting it upward for the Kon-Tiki, the second method lessens the amount of vapor created. The biochar produced was subjected to proximate characterization. Ultimate and heating value analyses, as well as the characteristics of the pore structure, were also evaluated. On the other hand, biochar has been field tested and applied to a bean crop at a rate of 1%. The harvested plants were tested for phytotoxicity, and parameters such as organic matter, pH, electrical conductivity, and cation exchange capacity were evaluated.

The biochar produced by TLUD has been found to exhibit somewhat better qualities in the parameters studied. This is supported by the field investigations, which show that the soils treated with this biochar had greater percentages of organic matter and more favorable pH values. Because Kon-Tiki's carbon and fixed carbon levels were somewhat lower than TLUD's, there was less organic matter present in the sample and more electrical conductivity. Notwithstanding these variations, both forms of biochar offer the best circumstances for field application as enhancers of the physicochemical characteristics of soil, hence promoting plant health. The application of biochar to the soil produced results that are consistent with those reported by other authors, who discovered that biochar enhances the physicochemical characteristics of the soil, reduces the bioavailability of potential inorganic contaminants, and facilitates plant uptake of essential nutrients (Acosta-Luque et al., 2022; López & Saldarriaga, 2023; Serrano et al., 2024).

Two inexpensive technologies that were tested on post-harvest biomass leftovers in low-income nations have both shown promise in the production of biochar. It has been shown that the physicochemical qualities of soil can be improved by this biochar; Kon-Tiki raises pH and EC, while TLUD improves CEC and OM. Additionally, the application of both devices' biochar greatly increased the foliar nutritional content of plants, demonstrating their feasibility for incorporating circular economy systems in agriculture by using leftover biomass to make biochar. An economic study shows that the first year's cost of implementing Kon-Tiki is more than €900 greater than that of TLUD. With bio-inputs made from leftover biomass, TLUD becomes a feasible choice for farmers looking to use circular economy techniques while improving soil qualities. Subsequent investigations ought to examine the effects of biochar on crops, specifically its capacity to augment pesticide retention, curtail application, and raise biodegradability.

References

Acosta-Luque, M., López, J., Henao, N., Zapata, D., Giraldo, J., & Saldarriaga, J. (2022). Remediation of Pb-contaminated soil using Biochar-Based Slow- Release P Fertilizer and biomonitoring employing bioindicators. Scientific Reports. https://doi.org/10.21203/RS.3.RS-2289314/V1

López, J. E., & Saldarriaga, J. F. (2023). Surface Co-application of Dolomitic Lime with Either Biochar or Compost Changes the Fractionation of Cd in the Soil and Its Uptake by Cacao Seedlings. Journal of Soil Science and Plant Nutrition, 1, 1–11. https://doi.org/10.1007/S42729-023-01469-Z/FIGURES/5

Serrano, M. F., López, J. E., Henao, N., & Saldarriaga, J. F. (2024). Phosphorus-Loaded Biochar-Assisted Phytoremediation to Immobilize Cadmium, Chromium, and Lead in Soils. ACS Omega. https://doi.org/10.1021/ACSOMEGA.3C07433

173am Enhancing Water Security and Defense: Exploring Graphene Nanoplatelet (GnPs) for Efficient Cyanotoxin Removal

Jesse Roberts

The escalating frequency of Harmful Algal Blooms (HABs) poses a significant concern, not only impairing the operational capabilities of the United States Army, but also presenting serious health risks due to the release of toxins into water sources. These toxins can adversely affect both military personnel and the civilian populace, underscoring the critical need for state-of-the-art water purification technologies. These cyanotoxins are not effectively removed by traditional treatment methods, such filtration and UV disinfection. Graphene nanomaterials have recently emerged as a promising solution for efficient and sustainable removal of emerging contaminants from water due to their high surface area, cost effectiveness, and ability to adsorb many classes of contaminants. Addressing this challenge is crucial for ensuring the health and readiness of our forces as well as protecting the health and safety of communities and ecosystems impacted by these blooms.

In this study, we demonstrate the use of graphene nanoplatelets (GnPs) as an adsorptive nanomaterial for the enhanced removal of cyanotoxins, including microcystin-LR (MC-LR), Anatoxin-a (ATX), and saxitoxin (STX). Using batch adsorption experiments, we benchmarked GnPs against the commercially available granular activated carbon (GAC) to evaluate the efficiency of GnPs for adsorbing these contaminants from water.

Our updated experimental results have shown that the sp2 carbon network and pore characteristics (pore size and morphology) of graphene make GnPs a highly effective adsorbent for the removal of all three cyanotoxins from water. GnPs, characterized by their mesoporosity and robust π-π electron coupling with aromatic toxin substrates, exhibit unprecedented adsorption capacities—133, 8, and 15 times greater for MC-LR, ATX-a, and STX, respectively, compared to conventional GAC. Sorption kinetic studies proved **graphene adsorbs 99% of MC-LR, ATA, and STX with pseudo-second-order kinetics in less than 30 minutes.** We provided mechanistic insight on graphene's enhanced ability to adsorb MC-LR by modeling the pore diffusion using intra-particle diffusion methods, calculating binding energies from DFT computations, and using NMR titrations to experimentally verify p-pi interactions between the aromatic ADDA chain of MC-LR and graphene's sp2 carbon network.

To gain deeper insight into the adsorption mechanisms, we fit 12 isotherm models to our data, elucidating the π-π , electrostatic, and pore diffusion interactions. Furthermore, we validated our findings with real-world spiked samples, confirming that no unwanted effects on adsorption were observed. Additionally, we expanded our study to include the removal of emerging contaminants and micropollutants, such as acetaminophen, caffeine, DEET, and DEP, demonstrating the versatility and broad-spectrum efficacy of GnPs.

Our conclusions prove the potential of GnPs in transforming water purification strategies, offering robust, rapid, and high-capacity solutions essential for military applications where water quality is paramount. The application of GnPs not only enhances the safety and efficiency of military operations but also holds considerable promise for civilian use, addressing public health concerns and protecting ecosystems from the detrimental effects of cyanotoxins and micropollutants. This pioneering research aligns with the defense sector's aim to innovate for environmental protection, providing a scalable and effective approach to managing water quality in diverse and challenging environments.

638a Strategizing Sorbent Design: Enhancing Removal of Emerging Contaminants for Water Remediation

Soyoung Choi, Kyriakos Stylianou, Ning Zhang, Aaron Moment

Emerging contaminants, including pharmaceuticals, personal care products, and herbicide, pose a threat to water quality and ecosystem health, including human society as they are not effectively removed with traditional wastewater treatment plants. The complex interactions between the adsorbents and emerging contaminants in various aqueous phase are still not yet fully understand, and hinder the strategical development of the materials for their enhanced and selective removal. This study, therefore, aims to address this challenge by exploring strategical approaches to design the engineered material via understanding the adsorption mechanisms and kinetics of selected emerging contaminants under various aqueous conditions. Here, we synthesized a water-robust MOF, UiO-66, and physically and chemically modified UiO-66 to determine the effective approaches to remove emerging contaminants. This study tested tetracycline, sulfathiazole, and atrazine as the representative emerging contaminants from the agriculture and livestock industry. Our findings reveal that physically modified MOF with controlled defects exhibited enhanced kinetics and isotherm relative to chemically modified adsorbents, even with presence of different types of co-existing cations (K, Na, Mg, and Ca) and anions (nitrate, sulfate) for tetracycline. Moreover, this physically modified MOF demonstrated robustness after multiple cycles of re-use, and exhibited enhanced selectivity towards hydrophilic emerging contaminant relative to activated carbon, a widely used adsorbents for water remediation. The results of this study offer valuable insights into the development of future sorbents with controllable properties for selective and enhanced water remediation, thereby contributing to addressing challenges in water-energy-food-environmental (WEFE) nexus.

638b Respirator Cartridge Performance Testing for Chemicals of Potential Concern (COPC's)

Satish Nune, Michael Minette, Angela Melville, Stephen, D Davidson, Luke El Khoury, Jasmine Gacutan, Thomas Brouns, Michael Zabel, Eugene Morrey, Kevin D. Miller, Chandra Lindberg, Margarat Bowman, Kenneth J Way

To accurately assess the performance of respiratory protection cartridges under varying environmental conditions, Pacific Northwest National Laboratory (PNNL) and Washington River Protection Solutions (WRPS) have jointly developed a laboratory test system to test respirator cartridges commonly used at Hanford tank farms in Richland, Washington, USA. The goal of this program is to understand respirator effectiveness across a wide range of vapors/gases and concentrations at wide range of humidities (50-80 RH) and high temperature (100 F). The Hanford Tank Farms has over 170 nuclear waste tanks that have varying vapor mixtures. For workers operating around those tanks, cartridge respirators are used to reduce potential health risks associated with chemicals of potential concern (COPC) in complex vapor mixtures. Laboratory testing will provide industrial hygiene professionals an understanding of the performance of respirator cartridges with these varying mixtures of COPC vapors under a range of environmental conditions. We will present our recent results on the experiments where the performance of the MSA OptiFilter TL multipurpose respirator cartridge was tested against a gas mixture containing tank waste representative gases that include furans, nitrosamines, aldehydes, ammonia, mercury, and nitrous oxide with varying chemical vapor concentrations. The tank waste representative breakthrough time can be used for establishing improved cartridge change out schedules and are improve worker safety and operating conditions. Laboratory results will be compared with the field results obtained through testing with actual tank vapors to benchmark the laboratory methodology.

638d Predicting the Photodegradation of Contaminants of Emerging Concern in Aquatic Systems

Emad Sanei, Itzel Marquez

Over the past few decades, there has been a growing acknowledgment of the ecological risks associated with the presence of contaminants of emerging concern (CECs) in aquatic systems, including pharmaceuticals, personal care products, and industrial chemicals. These compounds may enter the aquatic system due to not being fully removed from wastewater effluent. Understanding the fate of these substances upon their release into aquatic environments is crucial to contribute to the mitigation of their environmental impact. Photodegradation plays a significant role in CECs fate in the environment. CECs can undergo indirect photodegradation through reactions with reactive intermediates (RIs), including triplet-state organic matter ($^3DOM^*$), singlet oxygen ($^1O_2$), and hydroxyl radical ($^{\bullet}OH$), generated when waterborne organic matter is exposed to light. Organic material found in aquatic environments can originate naturally as natural organic matter (NOM) or enter the water system from wastewater treatment effluent as effluent organic matter (EfOM).

Characterization of organic matter is essential to understand RIs formation and, therefore, photodegradation mechanisms of CECs. However, characterization methods can be costly and complex. In an effort to simplify these methods, various optical parameters have been established to gain insights into the composition and photoreactivity of organic matter from UV-Vis and fluorescence spectroscopy. Indicators such as $SUVA_{254}$, HIX, FI, $S_{275-295}$, and E2:E3 are utilized to characterize EfOM in terms of molecular weight, aromaticity, and light absorption. Consequently, optical parameters of organic matter offer potential indications of its photoreactivity. This study developed a predictive method for the photodegradation of CECs by leveraging optical parameters of EfOM solutions as inputs.

Optical parameters including $SUVA_{254}$, HIX, FI, $S_{275-295}$, and E2:E3 for various EfOM solutions were measured. The EfOM solutions spiked with two CECs, *p*-cresol and propranolol, were exposed to simulated sunlight and contaminant degradation was determined. We then investigated relationships between optical parameters and kinetic parameters for the photodegradation of CECs using Pearson correlation. Using the correlations found, we predicted the degradation of CECs by solving the kinetic equation of degradation of each CEC. The prediction method was validated using different initial concentrations of CECs and several EfOM samples. The results demonstrated a favorable alignment between the experimental and predicted values.

By predicting the photodegradation of CECs, researchers and policymakers can assess the effectiveness of different treatment methods to minimize CECs persistence in aquatic environments. Additionally, accurate predictions of CECs photodegradation can inform risk assessments and regulatory decisions aimed at protecting human health and ecological integrity. This innovative approach not only enhances our understanding of CECs photodegradation mechanisms in complex EfOM solutions but also offers a practical tool for predicting CECs degradation kinetics, thereby facilitating the development of effective strategies for water treatment and environmental management. Moreover, it provides insights for designing effective monitoring programs, implementing targeted remediation efforts, and developing regulatory measures to prevent or mitigate CECs pollution. Furthermore, predictive models can assist in assessing the ecological risks associated with CECs exposure to aquatic organisms and ecosystems, guiding conservation efforts and ecosystem management strategies. Overall, the prediction of CECs in aquatic systems contributes to informed decision-making and proactive measures aimed at protecting water quality, biodiversity, and ecosystem health.

638e Investigation of Micropollutants Adsorption with Graphene Utilizing Interparticle Diffusion

Audie Thompson, Sarah Grace Zetterholm, Summer Nash, Ashvin Fernando, Chris Griggs

Micropollutants (MPs) are emerging as persistent contaminants that are recalcitrant to biodegradation and oxidation and can permeate reverse osmosis. Therefore, novel adsorptive media for tertiary treatment are required for safe effluent discharge. Interactions such as π-π and electron donor-acceptor with graphene can be a factor in the rapid kinetic removal of MPs. However, mechanistic insight is generalized, and solute characteristics are inadequately addressed. Here, we show the comparative adsorption of four aromatic MPs, caffeine, DEET, acetaminophen (APAP), and diethyl phthalate (DEP), with granular activated carbon (GAC) and graphene nanoplatelets (GnP). To understand the difference between the performance of GnP and GAC, the mechanism of removal was determined using experimental techniques, K_{ow} analysis, IPD kinetics, and NMR. Langmuir isotherm model correlation coefficients for GnP (R^2 0.36-0.99) were compared to the Freundlich isotherm correlation coefficients of R^2 of 0.44 to 0.99. All MPs showed Langmuir adsorption behavior, indicating monolayer adsorption, except DEET treated with GnP, indicating multilayer adsorption. When comparing the performance of GnP to GAC, GAC performed slightly better in most cases except for DEP. The capacity ratio of qe,max for GnP/GAC, was 0.866 for caffeine, 0.719 for APAP, 0.560 for DEET, and 1.002 for DEP. A comparison of the capacities of the sorbents revealed that GAC capacity was less dependent on the adsorbate, with a capacity range of 17.7mg/g, relative to GnP, which had a difference of 82.1 mg/g between DEP and DEET. This suggests the role of molecular structure in GnP's sorption capacity. As expected, GnP showed significantly faster kinetics for all MPs with k_2 values 2-3 orders of magnitude higher than GAC. However, while GAC capacity exhibited insensitivity to molecular structure, GnP adsorption capacities were influenced by the solutes' structure, with DEP > Caffeine > APAP > DEET in terms of capacity. For all intraparticle diffusion Kp, two values were higher for GAC except for DEP, which has the highest capacity for GnP. The data demonstrates that after the initial attraction, the MP must be able to diffuse into GnP's mesopores to have high adsorption capacity. This phenomenon is confirmed by NMR, which indicates that Kp,2 dictates capacity. Therefore, micropore diffusion has a greater effect on the adsorption of MPs for GAC than for GnP. The external mass transfer will remain rapid, but the capacity will be limited to surface interactions if structural constraints such as steric hindrance cause resistance to solute entry into the graphene particle. Overall, the results show that the performance is molecule- and structure-dependent. Other molecular structure considerations such as heteroaromaticity and planarity play a role in limiting intraparticle pore diffusion, thereby limiting overall capacity for GnP compared to GAC. Choosing the correct molecular structure to interact with GnP is pivotal in designing effective and rapid treatment solutions for the environmental treatment of emerging contaminants like MPs. Using the selectivity available with GnP can lead to improvements in capacity for treatment opposed to GAC in complex water matrices.

638f The Impact of Lead Water Pollution: Evidence from Three Case Histories

Robert Peters, Gail Wallace, Christobel Asiedu, Ghislain Gueye, Teresa Murray

Although water is considered as being essential to life and an important natural resource, disadvantaged communities such as low-income and minority communities are disproportionately burdened by lead exposure in drinking water. In this presentation, we highlight case studies that received national press coverage as well as recent examples of community lead poisoning hazards that are still ongoing across various regions of the U.S. We highlight three case studies of lead contamination involving Flint, Michigan, Washington D.C., and Birmingham, Alabama. Lead in water supplies causes a number of human health issues, including: developmental delay, learning difficulties, lower IQ, trouble concentrating, damage to the brain and central nervous system, slowed growth, irritability, loss of appetite, weight loss, fatigue, abdominal pain, vomiting, constipation, hearing loss, speech problems, decreased ability to pay attention, underperformance in school, headaches, seizures, among others. The public health and environmental toxicology implications of lead poisoning are described. Community grass roots organizing and partnering public health and environmental sustainability programs are needed to help communities address the cognitive and physical health implications of lead contaminated water and exposome threats that are lurking at disproportionately higher levels in impoverished racial and ethnic communities across the United States.

638g Assessment of Heavy Metals in Qatar's Groundwater: Implications for Water Quality and Human Health Impact

Mosab Subeh, Yehia Manawi, Jaber Al-Marri, Huda Al-Sulaiti

The quality of drinking water in Qatar relies heavily on its groundwater. This study investigates the monitoring of heavy metals, their distribution, and potential health impacts, offering crucial insights for effective groundwater management and human health protection. From January 2020 to December 2023, an extensive monitoring program analyzed groundwater samples using advanced analytical techniques like inductively coupled plasma-mass spectrometry and Ion Chromatography. Emphasizing the importance of monitoring heavy metals such as lead, arsenic, cadmium, chromium, and mercury, the study highlights their potential entry into groundwater through natural processes or human activities. Concentrations of these heavy metals were evaluated against international drinking water standards to assess health risks. Examination of 87 wells revealed a hierarchy of heavy metal prevalence, with average concentrations ranking highest for Sr > Si > B > P > Mo > Ba > Sb > Se > Tl > U > V > Ag > Cu > Al > As > Fe > Li > Zn > Cr > Mn > Co > Ni > Pb > Th > Be > Cd. Notably, Boron, Molybdenum, and Strontium concentrations exceeded WHO guidelines, ranging from 0.05 to 4.15 ppm, 0.001 to 0.32 ppm, and 0.016 to 24.34 ppm respectively. Recognizing the significant risks posed by heavy metals and metalloids to human health, even at low concentrations, underscores the need for proactive measures. This research provides vital insights into groundwater resource management and environmental planning, facilitating strategies to prevent and mitigate heavy metal contamination. Furthermore, the study explores the chronic daily intake (CDI) of investigated heavy metals (Ag, Mn, Cr, V, Mo, and Sr) through ingestion and dermal pathways, revealing a range of 1.4 × 10^-5 - 6.7 × 10^-1 mg/kg/day. Additionally, the Nemerow comprehensive pollution index (NCPI) highlights the multifaceted nature of pollution, urging holistic mitigation approaches.

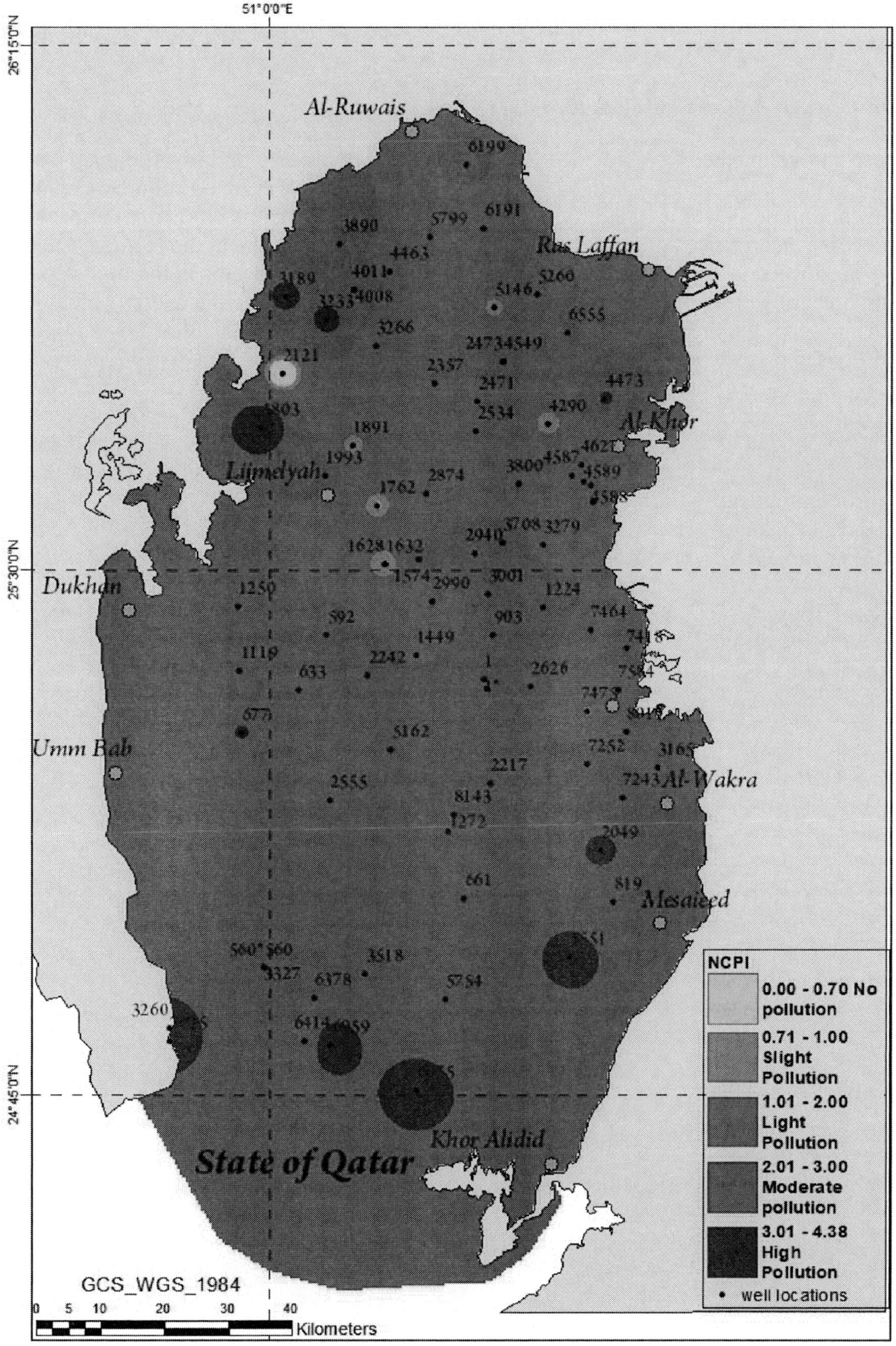
51°0'0"E
26°15'0"N
25°30'0"N
24°45'0"N
Al-Ruwais
Ras Laffan
Al-Khor
Dukhan
Umm Bab
Al-Wakra
Mesaieed
Khor Alidid
State of Qatar
GCS_WGS_1984
0 5 10 20 30 40
Kilometers
NCPI
0.00 - 0.70 No pollution
0.71 - 1.00 Slight Pollution
1.01 - 2.00 Light Pollution
2.01 - 3.00 Moderate pollution
3.01 - 4.38 High Pollution
well locations

261a Exergy Combined Economic Analysis As a Novel and Simple Energy Optimization Methodology: Application to Regeneration Process of Spent Rds Catalyst

Young-Hwan Chu, Dong-Hyuk Chun, Byung Hyun Kang

We propose a novel method to easily identify specific unit processes with low efficiency and high energy cost in the overall process using a specially devised exergy combined economic analysis. Once the inefficient unit processes are identified, then we can focus our efforts on them to improve the overall process performance and achieve a better energy-saving process. In the method, the exergy analysis is performed to investigate the exergy values of all process streams in the first step. Next, the capital and operating costs of each unit process are estimated to combine the exergy values obtained in the previous step. Finally, the production cost of each unit process is calculated by adding its input stream cost based on the exergy value to the capital and operating costs, and dividing it by the mass flowrate. By cost balance between input and output streams, high production cost is caused when exergy destruction is large or capital and operating costs are high. We applied the proposed method to a regeneration process of the spent RDS catalyst to verify its effectiveness. As a result, three roasters showed relatively high production costs despite their moderate exergy efficiency. The result means that it is difficult to identify the cost-intensive unit processes by exergy analysis alone. By rearranging the whole heat exchanger network through pinch analysis and optimally splitting oxalic acid solution feed stream to vanadium extractor into two separate streams, we could significantly reduce the production costs of the three roasters and achieved remarkable improvement in total energy consumption. Thus, the application result shows that the proposed method enables us to detect unit processes with low efficiency in terms of both exergy and cost and provides us with a new opportunity for process improvement and rationalization.

261b Sustainable Assessment of Emerging Technologies By Large-Language-Models-Based Genai Technology

Mahboubeh Moghadasi, Abdurrafay Siddiqui, Yinlun Huang

The next generation of technological competition in manufacturing industries will be dictated by inventions. Although emerging technologies can become an engine of change and progress, the net profit brought to the environment and society could be questionable, if sustainability principles are not fully incorporated into technology development phases. This renders a need for comprehensive sustainability assessment of technologies in their different development stages. In this endeavor, a key step is to identify the technologies at different technology readiness levels, ranging from the basic research level to the ready-for-adoption level. In the metal and polymeric coating and surface finishing sectors, there exists numerous technical information openly accessible. Identification and classification of technologies by conventional search methods are time consuming, inefficient, and susceptible to errors.

The realm of Generative AI (GenAI) is considered on the cusp of a new transformative leap, transitioning from Large Language Models (LLMs), which are multi-billion parameter transformer neural networks that are trained on enormous collections of documents without supervision or labels. LLMs can perform multiple tasks like classifying natural language, translating text, and document search. Perhaps the most remarkable task of LLMs is to complete an input string of text. Via this mechanism, LLMs can write unit tests, document function, write code from a doc string, answer questions, etc. The recent big leap in AI-based assistant chatbots, GPT-4, has created new opportunities to automate engineering tasks and reduce the workload on human experts. This work focuses on harnessing GPT-4 for the identification of emerging technologies for sustainability improvement of surface finishing. Our investigation delves into the capabilities of this advanced model in executing tasks related to technology identification and classification. We will share our experience in generating chain-of-thought and maieutic prompts, exploring the benefits, pitfalls, and challenges of using it. The sustainability assessment and readiness levels of a resulting set of technologies will be also presented to show when and how these technologies could be used in the manufacturing sector.

261d Establishing the Framework for a Process Systems Sustainability Index

Mitchell Huffman, Qingsheng Wang, Faisal Khan

As the human population grows and continues to industrialize, the strain on resources and the environment grows exponentially. Sustainable practices are therefore necessary to protect the planet as well as preserve the standard of living of those inhabiting it. As the process industry allows for such a large-scale impact on both people and the planet, the quantification of sustainability is vital to monitor and regulate industry practices. From publication data, it is also clear that there is a rising demand for various tools to measure the sustainability of a process. Therefore, it is necessary to understand the development of the field of sustainability quantification.

Frameworks such as life cycle assessment, impact assessment, and the triple bottom line provide approaches to quantify, understand, and implement sustainability concepts. Many studies have expanded on these frameworks to provide a variety of methods to quantify sustainability in industry, with various strengths and weaknesses for each. The key strengths, weaknesses, and resulting opportunities of these various methodologies will be discussed. To determine the validity of existing metrics, the consistency of their results must be compared. When utilizing existing sustainability indices, significant variation was found both in the final quantification of sustainability for a single process, and in the determination of a more sustainable process when comparing two processes. These inconsistencies imply the need for additional considerations within the various indices. From the analysis of existing indices, it was determined that sustainability metrics for process industry sustainability to this point have lacked an inclusion of safety, time dependence, and quantification of the interrelations between sustainability pillars.

As process safety incidents are so intricately tied to process economic feasibility, environmental impacts, and societal acceptance, it is determined that process safety is a key aspect of process sustainability that has been disregarded until recently. This work proposes that process safety is on par with the economic, environmental, and societal aspects of sustainability for the process industry, and therefore sustainability metrics in the area should expand the triple bottom line for the basis of their methodologies. Additionally, when a process is impacted in one of these four key aspects of sustainability, the impact on other aspects should be quantified. In the case of a process safety incident, key indicators in the economic, environmental, and societal sustainability of a process will be impacted, and any sustainability metric should reflect these cascading effects.

This work proposes a framework to establish a sustainability index for the process industry. This framework utilizes an expanded triple bottom line (environmental, economic, societal, safety pillars) and the driving force, pressure, state, exposure, effect, action (DPSEEA) method to establish and categorize indicators. Then, interrelations are established between the sustainability pillars within the expanded triple bottom line, and the interpretive structural modeling method is used to establish a hierarchy of dependence and influence of the different indicators on one another. This framework establishes a foundation that will be expanded upon to quantify sustainability within the process industry.

590a Evaluating the Electrochemical Operating Window for Titanium Metal Reduction Via Molten Oxide Electrolysis

Matthew Watson, Kathryn Ford, Rebecca Newport, Aaron Marshall, Catherine Bishop

Problem:

Molten oxide electrolysis (MOE) could enable a low-carbon route for sustainable production of metals. All-oxide MOE does not create toxic halide waste, can use electrons generated from renewable energy sources and will not directly produce greenhouse gases if inert anodes are used. MOE has significant technical challenges including electrolyte chemistry, ionic conductivity, and stability; crucible and electrode stability, conductivity, and longevity; and feedstock chemistry.

Titanium was selected as the metal of interest because it has a high strength-to-weight ratio, excellent corrosion resistance, and a high melting temperature. Despite the abundance of Ti-bearing ores, Ti is not widely used due to the expense and complexity of the incumbent production method – the Kroll process. [1] For this reason, we investigated a MOE route to Ti production.

To build a fundamental understanding of the electrolytic operating window a simple binary oxide system was studied. In this work sodia (Na_2O) was experimentally investigated as the supporting electrolyte for titanium oxide reduction to metal. Sodia dissolves oxides of titanium, and the TiO_2-Na_2O pseudo-binary system forms a eutectic, reducing the liquidus temperature over a wide composition range (Figure 1).

Methods:

The FToxid and FactPS databases of FactSage 8.1 [2] were used to make thermodynamic predictions, based on the real solution (i.e. non-ideal) nature of the TiO_2-Na_2O system. The predictions show that a reordering of reduction potentials occurs around the congruently melting line compound (Figure 2). In a reducing (argon + carbon) atmosphere, the predicted complete Ti^{4+}/Ti and partial Ti^{4+}/Ti^{3+} reduction potentials were less negative than the Na+/Na reduction potential at 1100°C and . Therefore the TiO_2-Na_2O system is predicted to be suitable for MOE to produce Ti metal.

To confirm the prediction, molten oxide electrolysis was carried out in a vertical tube furnace. Samples were placed in the centre of the hot zone using an internal support structure made of molybdenum (Figure 3) which was contained in an alumina tube inside the vertical tube furnace. Graphite crucibles were used both to contain the sample and to prevent any damage in case of leakage. The alumina tube was purged with argon to prevent oxidation of the internal components. An electrolysis experiment was performed at 1102 ± 5°C on an electrolyte with mol fraction.

After the electrolysis experiment, the furnace was cooled and a cross section of the now solid oxide mixture with the working electrodes still embedded was prepared. The working electrode was analysed via scanning electron microscopy (SEM) and elemental energy-dispersive X-ray spectroscopy (EDS) to determine the extent to which Ti metal had been deposited.

Results:

Ti metal was detected at the platinum working electrode (Figure 4) in a 20 ± 3 μm wide interdiffusion zone (region II) after electrolysis of the melt for 160 minutes at -0.25 V vs Ti reference. Up to 10 ± 1 wt% Ti was identified in the interdiffusion zone, which exceeds the equilibrium solubility of ~2.7 wt% Ti in Pt at 1100°C, [5] hence these are the likely intermetallic phases present in the WE interdiffusion zone (region II).

Ti reduction efficiency was estimated from 10 line-scans around the working electrode cross section. Electrolytic reduction of Ti was confirmed; however, the calculated reduction efficiency for Ti metal was very low ($0.06 \pm 0.02\%$). Decreasing the reduction voltage from -0.25 to -0.1 V vs the Ti reference electrode resulted in a reduction efficiency for Ti metal of $0.24 \pm 0.08\%$ for . Hence Ti reduction from TiO_2-Na_2O under these conditions is not practically viable.

The low reduction efficiency could be due to co-reduction of sodium. The experiment was performed at mol fraction and the compositional uncertainty was the difference between the mixed precursor (0.54) and XRF analysis of the product material (0.62), which indicated a significant amount of Na vaporisation occurred. Sodium metal was not identified at the electrode because the boiling point is 883°C. Therefore, any sodium metal produced at ~1100°C evaporated and was carried away by the argon.

Implications:

Thermodynamics calculations accounting for the non-ideality of the TiO_2-Na_2O pseudo-binary system correctly predicted that Ti metal would be electrolytically reduced over Na metal under certain conditions. Ti was observed alloying into the Pt working electrode for the experiment conducted at mol fraction. However, despite Ti reduction being confirmed the reduction efficiency was very low (below 0.25%).

While Na_2O is an unsuitable electrolyte for Ti reduction via MOE, we speculate our work forms the basis for a systematised approach to search for supporting all-oxide electrolytes. The supporting electrolyte must fulfil established criteria [6] such as having: the ability to dissolve the metal of interest; high ionic conductivity; and chemical and material [7] stability. In addition, in our work we looked for a deep eutectic leading to a substantial reduction of the liquidus temperature over a wide range of composition for more readily accessible MOE operating temperatures. Furthermore, we hypothesize that searching for oxide mixtures with a congruently melting line compound which leads to a reordering of reduction potentials can aid the discovery of all oxide mixtures suitable for MOE. However, this approach is limited by a paucity of high temperature data for binary and terniary oxide mixtures and a lack of predictive tools.

This research is funded by New Zealand Ministry of Business, Innovation and Employment (MBIE) contract CONT-76356-ENDSI-UOC.

References:

1. Kroll, W. (1940). The production of ductile titanium. *Transactions of the Electrochemical Society*, *78*(1), 35.

2. Bale, C. W., Bélisle, E., Chartrand, P., Decterov, S. A., Eriksson, G., Gheribi, A. E., ... & Van Ende, M. A. (2016). Reprint of: FactSage thermochemical software and databases, 2010–2016. *Calphad*, *55*, 1-19.

3. Hughes, T. G. (2018). The development of ultra-high temperature experimental capabilities for the electrolytic extraction of titanium from New Zealand steel's iron slag. *Thesis, University of Canterbury.*

4. Martin-Treceno, S., Hughes, T., Weaver, N., Marshall, A. T., Watson, M. J., & Bishop, C. M. (2021). Electrochemical study on the reduction of Si and Ti from molten TiO2–SiO2–Al2O3–MgO–CaO slag. *Journal of The Electrochemical Society*, *168*(6), 062502.

5. Biggs, T., Cornish, L. A., Witcomb, M. J., & Cortie, M. B. (2004). Revised phase diagram for the Pt–Ti system from 30 to 60 at.% platinum. *Journal of alloys and compounds*, *375*(1-2), 120-127.

6. Allanore, A. (2014). Features and challenges of molten oxide electrolytes for metal extraction. *Journal of The Electrochemical Society*, *162*(1), E13.

7. Ford, K. T., Marshall, A. T., Watson, M. J., & Bishop, C. M. (2024). Experimental Validation is Always Required for Molten Oxide Electrolysis Laboratory Crucibles. *Metallurgical and Materials Transactions*

B, 1-18.

Figure Captions:

Figure 1: FactSage 8.1 predicted TiO_2-Na_2O phase diagram at unit activity oxygen using the FToxid database. [2]

Figure 2: TiO2-Na2O reduction potentials at 1100°C for mol fraction . Lines are FactSage 8.1 predictions.

Figure 3: Electrochemical cell used in the vertical tube furnace, showing (A) the various individual components of the internal support structure system and (B) the modified area of the molybdenum working electrode in contact with the slag. Adapted from [3,4].

Figure 4: Analysis of the product at the Pt working electrode. a) EDS maps of Na, Ti and Pt, b) secondary electron image indicating line scan in yellow and c) EDS line scan across the interface between the oxide mixture and the Pt working electrode from b).

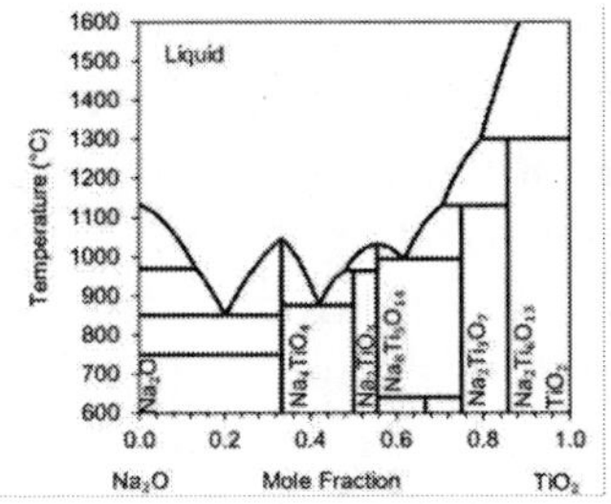

Figure 1

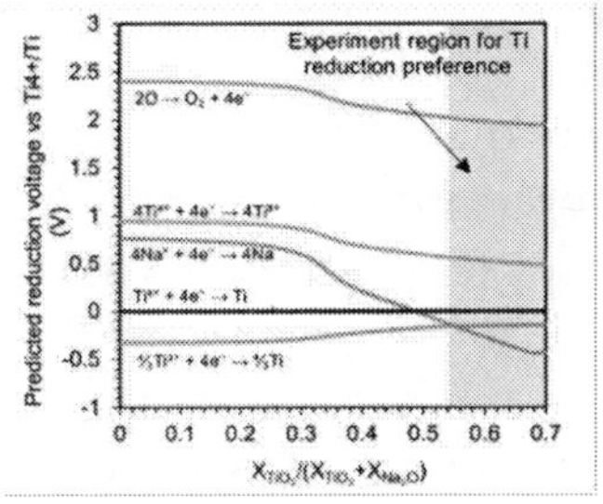

Figure 2

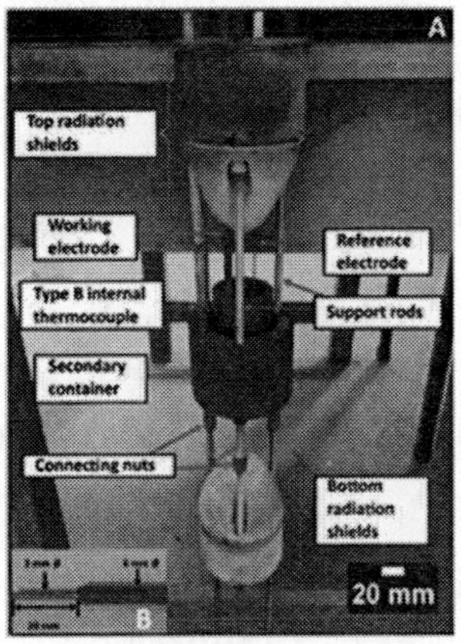

Figure 3

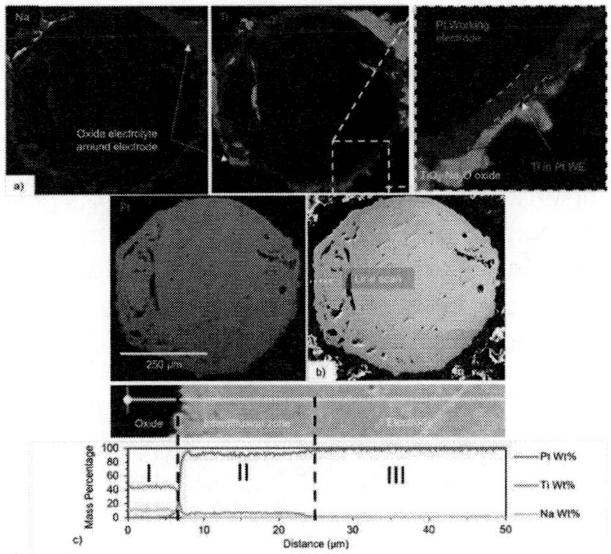

Figure 4

590b Controllability Evaluation of Intensified Separation Zones of a Carbon-Hydrogen-Oxygen Symbiosis Network

Maricruz Juárez-García, Juan Gabriel Segovia, Gabriel Contreras-Zarazúa, José María Ponce-Ortega

Carbon-Hydrogen-Oxygen symbiotic networks (CHOSYN) are settled for multi-plant mass integration among hydrocarbon processing plants as a strategy for reaching more sustainable industrial processes through performance targets such as reduction of raw material usage and waste generation, but also economic targets such as cost reductions and profitability enhancement [1]. Since these aspects are the guidelines used for the design of the networks, it is important to consider that the separation zones represent a significant contribution to the total cost and consume a large percentage of the energy requirements of the process, especially distillation processes that are well known for their low thermal efficiencies [2]. To address this concern, it has been proposed to include intensified distillation processes in the design of CHOSYNs as a strategy to improve the sustainability indicators of the networks such as economic and environmental aspects [3]. However, intensified distillation sequences such as thermally coupled sequences generally involve more complex designs than their conventional counterparts; therefore, in this work, it is proposed a controllability analysis performed in these intensified zones of the CHOSYN to determine whether the intensification improves or worsens its controllability. In this way, it is possible to ensure a feasible network design that incorporates the benefits of intensification and presents good controllability. Also, according to the principles of green chemistry by ensuring that the net design has good control properties a safer process can be expected. As a case study, it is proposed a CHOSYN configuration with two distillation sequences as areas for intensification [3]:

1. a) A three-column distillation sequence for ethylene purification from an ethane cracking process. For this sequence were proposed three retrofitting options replacing the reboiler from the first column (R-I option), the condenser of the second one (R-II option), or both (R-III option) for liquid/vapor recycles.
2. b) A two-column distillation sequence for propylene purification from methanol to propylene process. For this sequence were proposed two options: Petlyuk column and a side stripper arrangement.

The analysis was implemented under different scenarios with different intensification options considered for both sequences. The methodology implemented consists of two main analyses:

1) SVD method. First, the singular value decomposition technique was employed to assess the theoretical controllability of the different scenarios. This method allows us to measure qualitatively the controllability of a processing system through the concept of condition number, which is an indicator of how "well-conditioned" or sensitive is a matrix, in this case, this matrix is the steady state gain matrix, which relates the control and manipulated variables of the system. A large condition number indicates a sensitive system where a small disturbance in the system results in large changes in the control variables, conversely a small condition number indicates a less sensitive system. The matrix coefficients were determined by introducing a small disturbance in the manipulated variables and measuring the variation in the control variables, for this it was used Aspen plus simulations. For this criterion, the preliminary results indicate that some scenarios, which include intensification options, are as controllable as the conventional network. The intensified options for propylene purification present almost invariant condition numbers compared to the conventional sequence. Meanwhile, for ethylene purification, the intensified options presented higher condition numbers over

the conventional option, which means that control is negatively affected by these options.

2) A dynamic study. Conventional L/V control structures based on PI controllers for the different intensified distillation options were proposed. These control loops consist of a feedback control scheme where the purity of the dome and bottom products are controlled with the reflux ratio and the heat duty. Aspen Dynamics simulations were used to verify the dynamic response of the network for the different scenarios, a Set Point change of 1 % for the product purity was proposed and the minimization of the absolute error integral (IAE) was used to determine the values of the controller parameters Kc (proportional gain) and τi (integral time). The responses were compared with the non-intensified options. The results show that PI-based structures of control are adequate to control de proposed sequences, and the time of response for some intensified options does only slightly differ from the response of the conventional case.

The joint results of these two studies indicate that the complexities introduced by the intensification do not necessarily mean worse control of the network. Therefore it is possible to integrate intensified separation processes into the CHOSYN, thus improving the performance of the network in terms of sustainability, lower cost, and energy consumption, as well as guaranteeing a feasible design in terms of control. In this way, processing schemes such as CHOSYNs are emerging as tools for greening existing refinery and petrochemical processing systems.

[1] Noureldin, M. M. B., & El-Halwagi, M. M. (2015). Synthesis of C-H-O Symbiosis Networks. AIChE Journal, 61(4), 1242–1262. https://doi.org/10.1002/aic.14714

[2] Etchells, J. C. (2005). Process Intensification: Safety Pros and Cons, Process Saf. Environ. 83 (2), 85–89. https://doi.org/https://doi.org/10.1205/psep.04241

[3] Juárez-García, M., Contreras-Zarazúa, G., Segovia-Hernández, J. G., Ponce-Ortega, J. M. (2022). Sustainable Carbon–Hydrogen–Oxygen symbiosis networks: Intensifying separation sections. Chem. Eng. Process. - Process Intensif., 179, 109092.
https://doi.org/https://doi.org/10.1016/j.cep.2022.109092

590c A New Method for Lithium-Ion Battery Recycling Based on Chloroaluminate and Chloroaluminous Complexes

Nastaran Shojarazavi, Leo W. Gordon, Jian Zhang, Hossein Pazooki, Yuqing Fu, Chengi Hung, De-en Jiang, Jiayan Shi, Robert Messinger, Juchen Guo

The sustainable electrification of transport hinges on effective recycling strategies for spent lithium-ion batteries (LIBs), given the finite nature of critical metals like lithium, cobalt, and nickel. Conventional recycling techniques, primarily pyrometallurgy and hydrometallurgy, present limitations that hinder technological advancement in this field. In response, we propose a groundbreaking recycling method that circumvents existing barriers by leveraging the facile reaction between chloroaluminate (Al-Cl complex anions) and chloroaluminous (Al-Cl complex cations) species and layered lithium transition metal oxides (LTMOs).

Our results elucidate the robust performance of $AlCl_4^-$ in Lewis neutral chloroaluminate ionic liquid and $AlCl_2^+$ in aluminum chloride solution in ethanol ($AlCl_3$-EtOH) in chemical delithiation of LTMOs (including lithium cobalt oxide ($LiCoO_2$) and lithium nickel manganese cobalt oxide ($LiNi_{0.6}Mn_{0.2}Co_{0.2}O_2$)). The strong Li-Cl attraction initiates chemical extraction Li^+ cation from the LTMO that leads to a self-promoted reduction of transition metals to soluble lower oxidation states. This mechanism forms the basis of our innovative recycling methodology, which employs chloroaluminate and chloroaluminous complexes as highly efficient metal extraction reagents. Notably, this extraction process achieves high efficiency even under modest conditions, highlighting its potential for scalable implementation in recycling facilities.

Significantly, $AlCl_3$-EtOH demonstrates superior extraction efficiency under mild conditions, coupled with the recyclability of the ethanol after reaction by rotary evaporation and resulting in a blue mass which is soluble in water followed by chemical separation of highly pure cobalt hydroxide and lithium chloride. Furthermore, we showcase the applicability of our method by conducting experiments with spent real LIBs, utilizing $AlCl_3$-EtOH as the extraction reagent. Leading to more than 99.9% extraction efficiency for both cobalt and lithium demonstrated the viability and effectiveness of our approach in addressing the pressing need for sustainable LIB recycling. In addition to its effective extraction of valuable metals, this method facilitates the straightforward separation of graphite. Since graphite remains insoluble, it can be effortlessly filtered out prior to the chemical separation and purification of the resulting blue mass.

In conclusion, our study introduces a novel recycling paradigm that overcomes existing technological barriers, offering a sustainable solution for critical metal recovery from spent LIBs. The utilization of chloroaluminate and chloroaluminous complexes-based reagents, particularly $AlCl_3$-EtOH, demonstrates remarkable efficiency and practical feasibility, underscoring its potential to revolutionize the battery recycling industry and contribute to the long-term sustainability of transport electrification.

590d Impact of Reactive Conditions on Wte Ash for Recovery of Elements and Waste Management

Janhvi Trivedi, Marco J. Castaldi

The economic growth in society has spurred a substantial increase in metal production, notably in modern high-technology industries, where precious, critical, and rare earth elements (REEs) are in high demand. To overcome this high demand, currently the US imports about 95% of these elements from different countries. Moreover, the escalating production of these elements is accompanied by growing challenges related to the limited availability of natural resources and costs associated with their extraction. To address these challenges, it is crucial to prioritize reducing the depletion of natural resources and improving the utilization of elements from waste. The considerable accumulation of materials within waste presents an opportunity to retrieve essential elements crucial for industrial applications. Notably, these elements are abundant in the ash derived from Waste-to-Energy (WTE) facilities, providing an avenue for efficient resource recovery and seamless integration into the circular economy.

This research focuses on the treatment of WTE ash with waste from other streams like gypsum waste from the construction and demolition sector and spent Fluidized Catalytic Cracking catalyst (FCC) from chemical industries. WTE ash is thermally treated (up to 1500°C) with these additives in 15, 30, and 50 weight percentages in reactive conditions to investigate the transformation in the crucial elemental compounds and how they can be converted into more extractible forms.

Various characterizations like XRD, SEM/EDX, gas chromatography (evolved gas analysis), and thermal analysis, are utilized to study the changes in the elemental compounds at different temperature ranges, after their exposure to reactive environments. The introduction of the additives has already demonstrated the modifications in the transition elemental compounds. This work is focused on studying the changes in precious and REE compounds. The results showed the formation of different oxides, elemental compounds with aluminum or silicon, and alumino-silicate compounds by altering the oxidation state of the elements. These precious and REEs compounds formed are soluble in different acids which can lead to ease in their extraction via the acid leaching process. Thus, elements present in the WTE ash can be recovered by including gypsum and/ or spent FCC catalysts.

Furthermore, combining the equilibrium calculations with the characterization results showed the formation of various mineral compounds that impact the slagging behavior of the ash. The mineral compounds formed after the incorporation of the additives elevated the ash fusion temperatures (AFTs) by 100-200°C, leading to a reduction in the slagging of ash at low temperatures. This improvement is because of the formation of alumina-silicate compounds which have higher melting points than the original compounds in the ash causing an increase in the AFTs.

This presentation will show comparative results for several conditions and discuss possible restates, mechanisms leading to the transformations in elemental compounds and their oxidation states, and the gases evolved.

Innovative Multiproduct Biorefinery Design: Unlocking Value from Mexican Biomass Resources

Carlos Rodrigo Caceres-Barrera, Eduardo Sánchez-Ramírez,
Juan Gabriel Segovia-Hernández, Heriberto Alcocer-García
University of Guanajuato, Guanajuato, Mexico.

Abstract

Global agricultural activities produce approximately 5.9 billion tons of residues annually, which are often disposed of through open burning, contributing significantly to CO_2 emissions. However, these residues hold potential for transformation into high-value products within biorefineries. In Mexico, lignocellulosic biomass, particularly corn stover, is abundant and underutilized. This study investigates the conversion of such materials into valuable compounds, including levulinic acid, γ-valerolactone, furfural, and hydroxymethylfurfural. Process simulations were performed using Aspen Plus to attain product purities exceeding 98%. Optimization efforts focused on reducing costs, energy consumption, and environmental impact, evaluating four thermal integration strategies for energy-efficient distillation. Results indicate that two of these sequences enhance all indicators. Specifically, the thermally coupled sequence achieved reductions of 13.2% in Total Annual Cost (TAC), 22.7% in the Eco-indicator 99 (EI-99), and 27.9% in energy consumption, while the side-stripper configuration achieved savings of 8.9%, 16.3%, and 14.7%, respectively. These advancements in the purification stages contribute to enhancing the economic and environmental feasibility of biorefinery processes.

Keywords: biorefineries, sustainability, process optimization.

Introduction

Bioproducts are gaining momentum as a viable and sustainable alternative to traditional petroleum-based products, offering pathways to achieving several Sustainable Development Goals (SDGs).[1] These bioproducts provide a means of shifting away from fossil-based resources toward renewable feedstocks, which helps in reducing greenhouse gas emissions and reliance on non-renewable materials. However, their economic viability remains a significant challenge due to the typically low concentrations of desired compounds in the reactor effluent, which are often highly diluted in water. This dilution results in high energy demands, especially during the purification stage—one of the most energy-intensive and costly phases in bioproduct manufacturing.[2,3] Addressing these energy demands is crucial for improving the competitiveness of bioproducts in the market and advancing the principles of a circular economy.

To mitigate these challenges, process intensification techniques, such as thermal coupling, have demonstrated significant energy-saving potential, achieving reductions of up to 30% in energy consumption compared to conventional distillation configurations.[4] These techniques enable more efficient separation processes, which not only reduce energy use but also contribute to lower operational costs and environmental impacts. In addition, optimizing the design and operation of key equipment to minimize energy requirements further supports sustainable development by reducing carbon footprints and waste generation across the biorefinery's lifecycle. This study systematically evaluates a range of process intensification strategies with the goal of optimizing total cost, environmental impact, and energy consumption within a multiproduct biorefinery setting. Specifically, the biorefinery is designed to simultaneously produce four high-value-added bioproducts, each requiring high purity levels to meet market standards. By implementing advanced simulation tools and optimization frameworks, this research identifies configurations that maximize resource efficiency while minimizing environmental burdens. The research findings underscore the essential role of thermal coupling in enhancing biorefinery performance by achieving significant energy and cost savings. Furthermore, the study highlights that efficient bioproduct manufacturing supports the circular economy, as it promotes the valorization of biomass residues, optimizes resource utilization, and lowers ecological impacts. As biorefineries move toward widespread implementation, adopting process intensification strategies will be key to realizing their full potential as sustainable, economically feasible alternatives to conventional refining methods. This research thereby contributes not only to the advancement of sustainable production methodologies but also to the broader objective of establishing bioproducts as central to a low-carbon, resource-efficient future.

Methodology

Case Study.

In the work of Caceres et al.[5], a multiproduct biorefinery was designed to enable the simultaneous production of levulinic acid (LA), gamma-valerolactone (GVL), furfural (FF), and hydroxymethylfurfural (HMF). The process, illustrated in Figure 1, begins with biomass undergoing pretreatment in reactor R1 with a dilute acid solution, followed by neutralization in reactor R2. In filter F1, the pretreated biomass is separated into a liquid stream (hemicellulose) and a solid stream (cellulose and lignin). Subsequently, cellulose undergoes hydrolysis in reactor R3, producing lignin as a solid byproduct and yielding a mixture of sugars and sulfuric acid. This mixture is processed in column C1 to remove the acid, yielding glucose and xylose, which are directed in equal proportions (50% each) to reactors R4 and R6. In reactor R4, glucose is converted into levulinic acid (LA) and formic acid; the resulting stream is divided into two fractions, with 25% sent to column C6 for LA purification, and 75% sent to reactor R5 for gamma-valerolactone (GVL) production, subsequently purified in column C3. This 75/25 distribution was identified as the most sustainable approach in the above-mentioned work. Simultaneously, reactor R6 produces an aqueous mixture of furfural (FF) and hydroxymethylfurfural (HMF), which are purified in two distillation columns to obtain FF and HMF as final products. Despite the biorefinery's integrated design focused on resource efficiency, opportunities to further reduce energy consumption were identified. The conventional purification sequence for FF and HMF was found to have room for improvement, this area is highlighted at the bottom of Figure 1, leading to the exploration of alternative purification strategies.

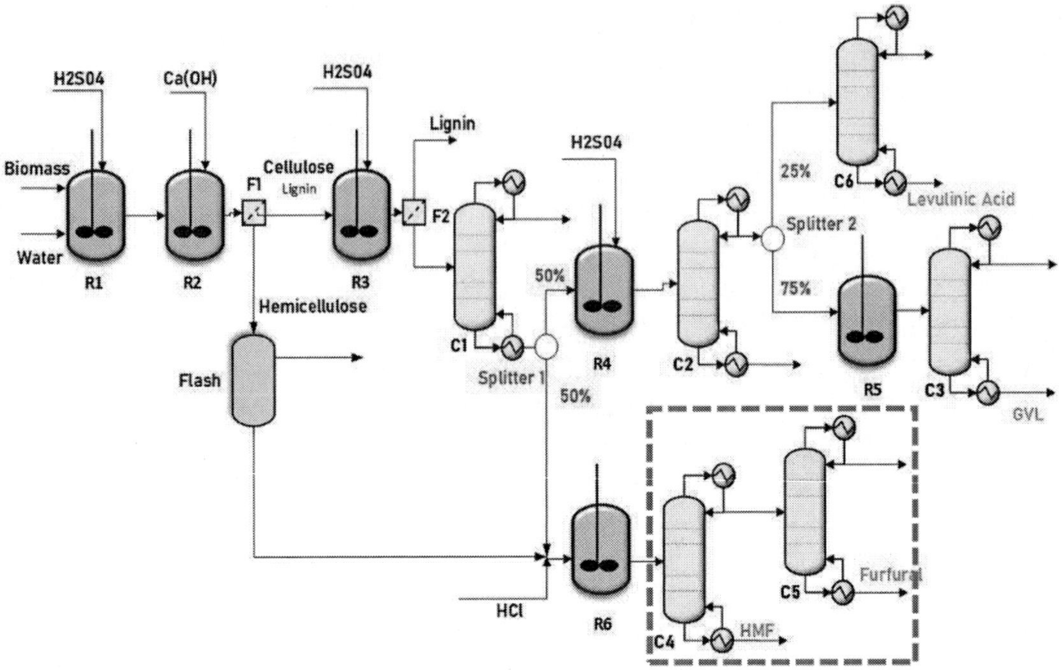

Figure 1. Multiproduct biorefinery for the case study. In red the distillation sequence to evaluate.

Design and Synthesis of Purification Alternatives

Starting with the base case of the conventional sequence (CS) for the purification of HMF and FF, consisting of columns C4 and C5 (Fig. 2a), several thermally coupled distillation sequence (TCDS) alternatives were evaluated to reduce energy consumption.[4,6] It is important to note that energy savings lead to indirect benefits, such as lower costs and reduced environmental impact. Alternative A is the indirect thermally coupled sequence (Fig. 2b), where the condenser of the first column is removed and replaced by a liquid stream from the second column, with a vapor stream also directed from the first to the second column. Alternative B is the side-stripper configuration (Fig. 2c), where sections of the setup are rearranged, and column C5 functions solely as a stripping section. Alternative C is the Petlyuk arrangement (Fig. 2d), which incorporates a prefractionation unit with interconnecting streams to the main column. Alternative D is the single-column arrangement with a side-product stream (Fig. 2e), where FF and HMF are obtained as bottom products.

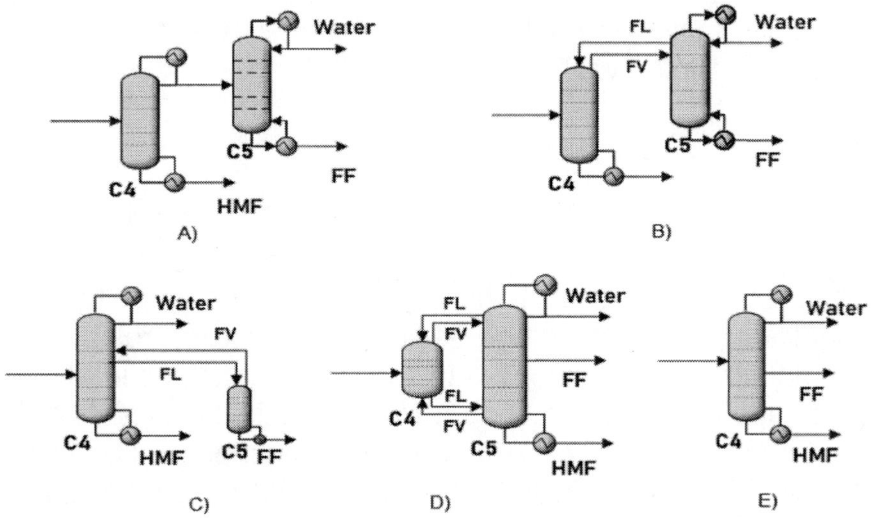

Figure 2. Different thermal couplings applied to columns C4 and C5

Optimization.

For optimization, three key sustainability indicators were used as objective functions. First, the total annual cost (TAC), calculated as the annualized equipment cost considering a 10-year return on investment plus utility costs. [7] Second, the Eco-indicator 99, which allows for the comparison of the environmental impact of different alternatives and was calculated following the methodology of Sanchez-Ramirez et al. [8] The third objective function is the total energy consumption of the process, derived from the simulation and expressed in MJ/year. Minimizing these objectives is challenging due to the high nonlinearity and nonconvexity of the model, so the differential evolution with tabu list (DETL) algorithm was chosen as the solution method.[9] This generation-based stochastic search method is well-suited for black-box models like process simulators, previous studies have demonstrated its effectiveness.[10–12].

The decision variables are the design variables for the biorefinery's distillation columns (C1 to C6). These variables are constrained within an operational range and include the number of stages (N), feed stage (Nf), reflux ratio (RR), distillate flow (D), column diameter (d), and, for coupled columns, the location (Fs) and values of the interconnection flows (Fi). Thus, the optimization problem can be formulated as:

$$Min(TAC, EI99, Energy) = f(N, Nf, RR, D, d, Fi, Fs)$$

The DETL implementation involves interaction between Microsoft Excel and Aspen Plus. The decision variable vector is sent from Excel to Aspen Plus, where the simulation is executed, and the resulting vector is returned to Excel. There, the objective functions are evaluated, and new decision variable values are generated. The parameters used were an initial generation of 120 individuals, 400 generations, a tabu list of 60 individuals, a tabu radius of 0.0001, and values of 0.9 for the crossover factor and 0.3 for the mutation factor. These values are taken from the literature, as parameter tuning was not considered[13].

Results

Table 1 presents the values of the objective functions for the optimal scenarios, with bolded figures indicating the percentage savings or increases relative to the base case. It is evident that thermal couplings A and B contribute to reductions in all performance indicators. Specifically, Alternative A have reductions of 11.4% in Total Annual Cost (TAC), 22.7% in the Eco-indicator 99 (ECO-99), and 27.9% in energy consumption. For Alternative B, the savings amount to 6.2% in TAC, 16.3% in ECO-99, and 14.7% in energy consumption. Conversely, Alternative C results in increases across all objective functions, with increments of 5.1% in TAC, 17.7% in ECO-99, and 21.9% in energy consumption. This outcome is further corroborated by Table 2, which shows that, despite an extensive search range for variables such as the reflux ratio and distillate flow, the optimization method was unable to yield superior results for the Petlyuk arrangement. The elevated energy demand stems from the arrangement converging only at high reflux values (18.4-18.5) to meet the required product purity, resulting in a substantial vapor flow rate of 12,600 kg/h and an increased reboiler load. Although Petlyuk columns typically reduce energy consumption for ternary mixtures with high concentrations of the lightest and heaviest components, the specific composition of this mixture limits the efficiency gains. Alternative D, while offering a 2.1% reduction in TAC, exhibits increases of 3.51% and 4.27% in ECO-99 and energy consumption. This is attributed to the savings in capital costs; however, the increased demand for steam raises both the eco-indicator and energy requirements. These results underscore the critical role of mixture composition in determining the effectiveness of thermally coupled sequences, as well as the importance of carefully balancing capital and operational costs to optimize sustainability and energy efficiency.

Table 1. Values for the objective functions for the different evaluated alternatives

Design	TAC (dollars/year)	Eco-99 (points/year)	Energy (MJ/year)
Conventional	$3.157 * 10^7$	$7.710 * 10^6$	$1.756 * 10^9$
Thermally coupled	$2.740 * 10^7$ **-13.2%**	$5.959 * 10^6$ **-22.7%**	$1.266 * 10^9$ **-27.9%**
Side Stripper	$2.875 * 10^7$ **-8.9%**	$6.452 * 10^6$ **-16.3%**	$1.497 * 10^9$ **-14.7%**
Single column	$3.093 * 10^7$ **-2.1%**	$7.981 * 10^6$ **3.51%**	$1.831 * 10^9$ **4.27%**
Petlyuk	$3.320 * 10^7$ **5.1%**	$9.08 * 10^6$ **17.7%**	$2.14 * 10^9$ **21.9%**

Table 2. Decision variable values for the optimal case of each scenario.

Variable	Conventional	Alternative A	Alternative B	Alternative C	Alternative D
Number of stages, C4	42	32	73	18	64
Feed stage, C4	27	4	44	17	63
Diameter, C4-C5, m	1.16	0.70	0.54	0.72	0.5
Flow C4, $\mathrm{kmol\ h^{-1}}$	94.94 (Distillate)	11.7 (Bottoms)	11.89 (Bottoms)	NA	11.89 (Bottoms)
Reflux ratio, C4	0.09	NA	7	NA	13
Liquid flow, C4 $\mathrm{kg\ h^{-1}}$	NA	72	1540	106.716	NA
Reboiler duty C4 (cal/s)	318637	303776	1.784×10^6	0	3.095×10^6
Condenser duty C4 (cal/s)	279991	0	-1.746×10^6	0	-3.053×10^6
Number of stages, C5	59	75	9	65	NA
Feed stage, C5	34	73	1	45 (Steam) 51 (Liquid)	NA
Diameter C5, m	0.72	0.637	0.5	0.57	NA
Flow C5, $\mathrm{kmol\ h^{-1}}$	80.74 (Distillate)	15.62 (Bottoms)	15.4 (Bottoms)	79.76 (Bottoms)	NA
Reflux ratio, C5	10.18	2.83	NA	18.95	NA
FV output stage	NA	NA	NA	46	NA
FL output stage	NA	72	NA	52	60
Flow FV $\mathrm{kg\ h^{-1}}$	NA	NA	NA	643.25	NA
Reboiler dutyC5(cal/s)	2.47×10^6	578543	92.6	4.42×10^6	NA
Condenser duty C5(cal/s)	-2.47×10^6	839551	0	-4.38×10^6	NA

Conclusions

Thermal coupling and optimization techniques were strategically applied to reduce total cost, environmental impact, and energy consumption within the framework of a multiproduct biorefinery. Among the four evaluated alternatives, two demonstrated notable improvements across these indicators. The thermally coupled sequence achieved significant savings, with reductions of 13.2% in Total Annual Cost (TAC), 22.7% in the Eco-indicator 99 (ECO-99), and 27.9% in energy consumption. Similarly, the side-stripper arrangement offered substantial improvements, with savings of 8.9% in TAC, 16.3% in ECO-99, and 14.7% in energy use.

In contrast, the Petlyuk column did not yield optimal results under the conditions tested, and, along with the single-column arrangement, failed to improve the proposed indicators effectively. The Petlyuk column's inefficiency in this case can be attributed to the specific composition and concentration distribution within the feed mixture, which limited the potential for energy savings typically associated with this configuration in ternary systems.

These findings highlight that reducing operational costs, energy demand, and environmental impact are key to enhancing the viability and competitiveness of biorefinery processes, particularly in comparison to traditional crude oil refining schemes. Such optimizations not only align with sustainable development objectives but also support the economic feasibility of biorefineries, positioning them as viable, eco-friendly alternatives for the production of high-value bioproducts. The insights gained from this study underscore the importance of selecting the appropriate thermal integration strategies tailored to feed composition and process specifications to achieve optimal performance in biorefinery systems.

References

(1) Arauzo J; Bimbela F; Abrego J. (2014), "Introduction to Biomass Valorisation Technologies". *Boletín del grupo español del carbón*, *33*, 2–6.

(2) Kiss, A. A.; Smith, R. (2020), "Rethinking Energy Use in Distillation Processes for a More Sustainable Chemical Industry." *Energy*, *203*, 117788. https://doi.org/10.1016/j.energy.2020.117788.

(3) Martín, M.; Grossmann, I. E. (2013), "On the Systematic Synthesis of Sustainable Biorefineries". *Ind Eng Chem Res* , *52* (9), 3044–3064. https://doi.org/10.1021/IE2030213.

(4) Segovia-Hernández, J. G.; Bonilla-Petriciolet, A.; Salcedo-Estrada, L. I. (2006),"Dynamic Analysis of Thermally Coupled Distillation Sequences with Undirectional Flows for the Separation of Ternary Mixtures". *Korean Journal of Chemical Engineering*, *23* (5), 689–698. https://doi.org/10.1007/BF02705913.

(5) Caceres, C. R.; Sánchez-Ramirez, E.; Segovia-Hernández, J. G. (2024) "Design and Optimization of a Sustainable Process for the Transformation of Glucose into High Added Value Products"; pp 73–78. https://doi.org/10.1016/B978-0-443-28824-1.50013-2.

(6) Keigo M; Koichi I; Masaru N. (2012) "*Recent Advances in Internally Heat-Integrated Distillation Columns (HIDiC) for Sustainable Development*"; Vol. 45.

(7) Turton, R. ,; Bailie, R. C. ,; Whiting, W. B. ,; Shaeiwitz, J. A. (2008) "*Analysis, Synthesis and Design of Chemical Processes*". ; Pearson Education.

(8) Sánchez-Ramírez, E.; Quiroz-Ramírez, J. J.; Segovia-Hernández, J. G.; Hernández, S.; Bonilla-Petriciolet (2015) "A. Process Alternatives for Biobutanol Purification: Design and Optimization" *Ind Eng Chem Res*, *54* (1), 351–358. https://doi.org/10.1021/ie503975g.

(9) Srinivas, M.; Rangaiah, G. P. (2007) "Differential Evolution with Tabu List for Solving Nonlinear and Mixed-Integer Nonlinear Programming Problems". *Ind Eng Chem Res*, *46* (22), 7126–7135. https://doi.org/10.1021/ie070007q.

(10) Alcocer-García, H.; Segovia-Hernández, J. G.; Prado-Rubio, O. A.; Sánchez-Ramírez, E.; Quiroz-Ramírez, J. J.(2019) "Multi-Objective Optimization of Intensified Processes for the Purification of Levulinic Acid Involving Economic and Environmental Objectives". *Chemical Engineering and Processing - Process Intensification*, *136*, 123–137. https://doi.org/10.1016/j.cep.2019.01.008.

(11) Contreras-Zarazúa, G.; Martin-Martin, M.; Sánchez-Ramirez, E.; Segovia-Hernández, J. G. (2022) "Furfural Production from Agricultural Residues Using Different Intensified Separation and Pretreatment Alternatives. Economic and Environmental Assessment". *Chemical Engineering and Processing - Process Intensification* , *171* (February 2021). https://doi.org/10.1016/j.cep.2021.108569.

(12) Romero-García, A. G.; Mora-Morales, C.; Chargoy-Amador, J. P.; Ramírez-Corona, N.; Sánchez-Ramírez, E.; Segovia-Hernández, J. G. (2022) "Implementing CO2 Capture Process in Power Plants: Optimization Procedure and Environmental Impact". *Chemical Engineering Research and Design*, *180*, 232–242. https://doi.org/10.1016/j.cherd.2022.02.023.

(13) Alcocer-García, H.; Segovia-Hernández, J. G.; Sánchez-Ramírez, E.; Caceres-Barrera, C. R.; Hernández, S.(2024) "Sequential Synthesis Methodology in the Design and Optimization of Sustainable Distillation Sequences for Levulinic Acid Purification". *Bioenergy Res*, *17* (3), 1724–1738. https://doi.org/10.1007/s12155-024-10765-0.

322a Microkinetic Modelling of Polyolefin Binary Mixture Pyrolysis

Aswathy K. Raghu, Paulami Majumdar, Linda J. Broadbelt

With about 400 million tonnes of plastic waste generated per year, plastic pollution is a major crisis currently facing humanity [1]. A significant portion of the plastic waste is made up of polyolefins such as polyethylenes and polypropylene. Pyrolysis is an effective chemical recycling strategy that is reasonably agnostic to the type of plastic feedstock and can therefore be applied to mixed streams. Pyrolysis of polyolefins results in hydrocarbons with low carbon numbers that can be used as fuels.

Since pyrolysis results in the formation of a wide range of products, detailed mechanistic models are needed to predict the product distribution. A microkinetic model is a multiscale model as it includes kinetic information of the mechanistic steps involved and predicts reactor level measurables such as conversion and product selectivity. The Broadbelt group has developed microkinetic models for pyrolysis of several polyolefins based on the population balance approach [2], [3], [4]. In this approach, the reaction mixture is divided into classes based on their characteristics that are relevant for the reactions (species types, end types, etc.), and the chain length distributions of these species classes are tracked using their moments. These models also have the capability to track explicit species up to specified carbon numbers for identification of the product distribution.

Polyolefin pyrolysis proceeds by a radical mechanism. When a mixture of plastics is pyrolyzed, they behave as a kinetically coupled system where the chain propagation steps are facilitated by the radicals from all the polymers. However, if the polymers in the mixture are immiscible, this crossover of radicals is not feasible for longer chains. In such situations, smaller radical species can still diffuse into other phases and facilitate chain propagation. This behavior affects the reaction time of each individual polymer in a mixture compared to what is observed in the pyrolysis of pure samples [5]. The rates of cross-propagation and self-propagation of radicals impact the depolymerization rates [6]. In this work, we present a microkinetic modelling study of binary mixture pyrolysis involving polypropylene (PP), high density polyethylene (HDPE) and linear low density polyethylene (LLDPE). PP/HDPE form an immiscible mixture whereas HDPE/LLDPE form a miscible mixture when LLDPE has a low branch content. Our preliminary results suggest that, in a PP/HDPE mixture, the degradation of HDPE is enhanced while PP is retarded in the mixture. A detailed analysis of the reaction mechanisms will be presented.

Figure 1: Comparison of decay trend of number average molecular weights of polyolefins during mixture and pure component pyrolysis of PP/HDPE mixture at 653.15 K.

References

[1] UNEP (United Nations Environment Programme), *From pollution to solution: a global assessment of marine litter and plastic pollution*, vol. 237, no. 3169. 2021. [Online]. Available: https://www.unep.org/resources/pollution-solution-global-assessment-marine-litter-and-plastic-pollution

[2] T. M. Kruse, O. Sang Woo, and L. J. Broadbelt, "Detailed mechanistic modeling of polymer degradation: application to polystyrene," *Chem Eng Sci*, vol. 56, no. 3, pp. 971–979, Feb. 2001, doi: 10.1016/S0009-2509(00)00312-2.

[3] T. M. Kruse, H. W. Wong, and L. J. Broadbelt, "Mechanistic modeling of polymer pyrolysis: Polypropylene," *Macromolecules*, vol. 36, no. 25, pp. 9594–9607, 2003, doi: 10.1021/ma030322y.

[4] S. E. Levine and L. J. Broadbelt, "Detailed mechanistic modeling of high-density polyethylene pyrolysis: Low molecular weight product evolution," *Polym Degrad Stab*, vol. 94, no. 5, pp. 810–822, 2009, doi:10.1016/j.polymdegradstab.2009.01.031.

[5] T. M. Kruse *et al.*, "Binary mixture pyrolysis of polypropylene and polystyrene: A modeling and experimental study," *J Anal Appl Pyrolysis*, vol. 73, no. 2, pp. 342–354, 2005, doi: 10.1016/j.jaap.2005.03.006.

[6] C. Lamarca, C. Libanati, and M. T. Klein, "Design of Kinetically Coupled Complex Reaction Systems," *Chem Eng Sci*, vol. 45, no. 8, pp. 2059–2065, 1990.

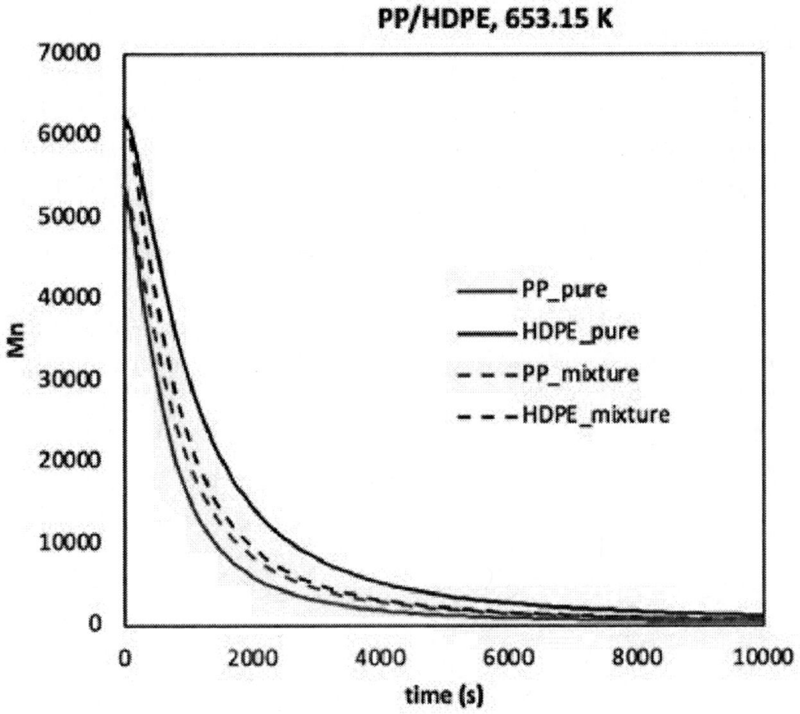

322b Green Solvents for Upcycling Waste PVC into Ultrafiltration Membranes for Water Treatment

Atta Ur Razzaq, Milad Esfahani

The increasing levels of plastic pollution have resulted in a growing threat to wildlife and ecosystems. Taking immediate action to stop this environmental threat and promote sustainable solutions is imperative. One step towards sustainability is to upcycle the plastic waste to high-value-added products such as membranes for water treatment. Non-solvent-induced phase separation (NIPS) using organic solvents is a standard process for membrane fabrication. These organic solvents, such as tetrahydrofuran (THF), N,N-dimethylformamide (DMF), pose environmental and health hazards as they are toxic and harmful to human health. To improve the eco-friendliness and sustainability of the waste plastic upcycling process, we have investigated the use of green solvents, such as Dihydrolevoglucosenone (Cyrene) and Triethyl phosphate (TEP), for the upcycling process of waste PVC pipes. These upcycled PVC pipe ultrafiltration membranes were thoroughly characterized to evaluate the impact of these two green solvents on membrane physicochemical properties. Each of these green solvents imparted distinct properties to the membrane. For instance, membranes prepared with Cyrene exhibited a sponge-like morphology, while those prepared with TEP showed a finger-like morphology. The performance of these upcycled membranes was also evaluated in terms of pure water permeability and Direct Red 80 (DR-80) dye rejection. The upcycled PVC pipe membranes exhibited impressive permeability (25 LMH/bar with Cyrene and 55 LMH/bar with TEP) and excellent DR-80 rejection (94.64% with Cyrene and 72.58% with TEP), demonstrating the significant potential for these green solvents for upcycling the waste PVC pipes into ultrafiltration membranes.

322c Electrochemically Mediated Alkaline Hydrolysis and Methanolysis of Poly(ethylene terephthalate) (PET)

Samantha Bunke, Kindle S. Williams, William A. Tarpeh

Background:

Decades of uncontrolled plastic production and disposal have resulted in large discharges of plastics into the environment and runaway ecological effects. The current plastic industry is unsustainable and hazardous, as it rapidly consumes finite resources and emits greenhouse gasses. Recycling plastic is an approach to mitigating these environmental effects; however, the most commonly used recycling processes employ mechanical methods of shredding and melting plastic waste, which degrades the polymers' properties and upholds an unsustainable, linear plastic economy. Chemical recycling offers a pathway to circularizing the plastic industry by utilizing reactive recycling processes to recover high-quality monomer feedstocks for repolymerization. Two such chemical recycling methods, alkaline hydrolysis and methanolysis, have been demonstrated to effectively depolymerize common plastics such as poly(ethylene terephthalate) (PET). However, both methods require environmentally intensive chemical feedstocks, as well as high energy inputs for extreme operating conditions (e.g., temperature and pressure), preventing their full-scale deployment. Our approach aims to reinvent chemical recycling by integrating it with electrochemical processes to reduce both costs and environmental impacts. We investigate electrochemically mediated alkaline hydrolysis and methanolysis to achieve the following advantages over conventional methods: access to more extreme reactivity from applying an electrochemical driving force, process intensification, integration with renewable energy, and application of more moderate operating conditions.

Methods:

Electrochemically mediated recycling converts plastic waste back into its monomer units by generating reactants electrochemically and depolymerizing plastic *in situ*. In a two-chamber cell separated by a cation-exchange membrane, both alkaline (pH>12) and acidic (pH < 2) environments are achieved via hydrogen evolution reaction (HER) in the cathode chamber and oxygen evolution reaction (OER) in the anode chamber, respectively. Conducting HER in a mixed solvent system of both methanol and water produces methoxide and hydroxide ions for use in depolymerization. The supporting electrolyte for both chambers is 0.1M $NaClO_4$, with a Pt coil as the cathode, an Ir/Ta mixed metal oxide mesh as the anode, and Ag/AgCl as a reference electrode. We targeted poly(ethylene terephthalate) (PET) for this approach because it is commercially available and a significant contributor to single-use plastic waste. Electrolysis was conducted at a constant current (20mA) for 5 hours and evaluated by three metrics: total PET conversion measured by mass difference in solids; the molecular weight distribution of the remaining oligomers measured by gel permeation chromatography (GPC); and the yield of monomer products, terephthalic acid (TPA), monomethyl terephthalate (MMT), and dimethyl terephthalate (DMT) measured by high-performance liquid chromatography (HPLC). We investigated the effects of several experimental conditions on the performance, including water/methanol ratio in the catholyte, catholyte temperature, and anolyte composition. While high PET conversion is one important goal, industrial feasibility of this recycling approach requires the selective production of either TPA or DMT for eventual repolymerization into high-quality plastic products.

Results and implications:

Our investigation of experimental conditions revealed several key insights for designing electrochemically mediated recycling. First, various performance regimes exist as a function of the catholyte water/methanol ratio, indicating that methanol plays a key role in both depolymerizing PET and improving mass transfer of reactants to PET. A minimum threshold of methanol content (≥30 mol% methanol) is needed to disrupt surface tension and allow for effective mixing of PET in the catholyte. In the 30-80 mol% methanol regime, there is a positive correlation between methanol content and total PET conversion. Interestingly, TPA selectivity also increases in this regime despite the decreasing water content, with water being the feedstock for the reactant that produces TPA. For the 100 mol% methanol case, selectivity shifts towards MMT and DMT, while total conversion decreases. TPA production still occurs in this case due to presence of a small amount of water that was transferred from the anode chamber throughout the 5-hour electrolysis period. Second, despite not being the direct reaction environment , the anolyte can also be engineered to improve overall performance. Using a buffer (Na3PO4) as the supporting electrolyte in the anode chamber can improve depolymerization by preventing the transfer of protons to the cathode chamber that would have otherwise neutralized methoxide and hydroxide ions. We observed >2x increase of PET conversion in the catholyte when a buffered anolyte was used. Additionally, when all considered experimental conditions were optimized, we achieved >60% PET conversion. The knowledge gained from studying electrochemically mediated depolymerization of PET will be applied to other common and novel plastic designs to expand the scope of this work and contribute to the movement towards a circular plastic economy.

322d Transient Thermal Barcode: A High Speed, High Throughput Plastic Sorting Based on Molecular Vibrational Signature.

Patatri Chakraborty, Yaoli Zhao, Thomas Thundat, Amit Goyal

While the production of plastic has increased exponentially over the years, from 2.3 million tons in 1950 to 448 million tons by 2015, the overall recycling rate of waste plastic has grown only to 8.7%. The first step towards efficient recycling is the accurate identification of each type of the 6 types of recyclable plastic. At present, the most effective sorting methods in the market are based on NIR, Raman spectroscopy & AI-assisted sorters. NIR spectra have overtones which increase the chances of false identification whereas Raman signals are very weak and require high-power lasers. AI-assisted sorting robots can sort materials based only on their visible characteristics, but cannot provide chemical information. Hence there is an urgent need to develop a novel technique which can sort plastic waste in a high throughput fashion that can be easily implemented in Material Recycling Facility (MRF). To address the abovementioned challenges, we have developed technique of plastic sorting called the Transient Thermal Barcode (TTB). The TTB technique is based on imaging localized heat generated by mid-IR absorption (photothermal effect). In the mid-Infrared region, called the 'molecular fingerprint regime' due to the absence of overtones, different plastics have unique spectral peaks. When IR with the matching wavelength is incident on the plastic, molecules are resonantly excited into higher vibrational states. Non-radiative decay of the excited states generates localized heating on the plastic. Projecting IR light patterns with multiple characteristic spectral peaks on the plastics generates photothermal heat patterns (Barcodes) characteristic to each of the 6 recyclable plastics including black plastic. We have demonstrated that imaging the plastic with an IR camera can identify different plastics based on their barcode heat pattern. This TTB technique has the potential for high throughput sorting of waste plastic in MRFs and secondary recycling facilities.

Acknowledgement: - This work was supported by the NYS-DEC Initiative for Plastics Recycling Research and Innovation, University at Buffalo. (Website: - https://www.buffalo.edu/iprri.html)

322e Biodegradation of Polymer Films and Fibers: Exploring the Role of Surface Coatings

Anicah O'Brien, Morton A. Barlaz, Saad A. Khan

The environmental devastation caused by single-use plastics necessitates the advancement of sustainable materials. Thus, materials that meet both in-use requirements and provide a viable alternative for sustainable disposal must be developed. Biodegradable plastics are a promising solution, as they can undergo biological recycling, where valuable products such as compost and biogas are produced upon disposal. Many biodegradable plastics are available on the market today, including (poly)-lactic acid, or PLA, a polyester derived from renewable biomass. Although PLA is certified as compostable at thermophilic temperatures, current research is predominantly focused on unmodified PLA. However, before integration into consumer products, PLA requires modification to overcome its hydrophobicity and brittleness. Yet, the impact of surface coatings, additives, and blending on PLA biodegradation remains elusive. In this study, we aim to unravel the impact of surface coatings on PLA biodegradation in relation to temperature, oxygen presence, and structure. PLA films and fibers were fabricated and coated with sorbitan monolaurate or polysorbate 20, chosen for their common use as wetting agents in consumer products. The materials were characterized through contact-angle measurements to assess the coating's impact on wettability and quartz-crystal microbalance to examine adsorption kinetics. Both aerobic and anaerobic biodegradability were studied as a function of temperature, as the material could be treated by composting (aerobic) or disposed of in a landfill or anaerobic digester (anaerobic) at the end-of-life. Carbon dioxide and methane generation were used to quantify aerobic and anaerobic biodegradation, respectively. Additional indicators of biodegradation were also measured throughout the duration of the experiments, including molecular weight, crystallinity, and biofilm thickness. The insights gained from this work will inform the development of sustainable materials with functional moieties for various applications including hygiene products, single-use packaging, and disposable medical devices.

322f Targeted Separation Scheme of Polyurethane Depolymerization Products

Taysha Telenar, Timothy E. Long, Matthew D. Green

In the United States, end of life polyurethane foams such as mattresses or cushions go predominantly to landfills. Mechanical recycling can prolong the inevitable fate of these materials, but not indefinitely. An emerging option is chemical recycling: breaking down these foams into smaller re-polymerizable feedstocks with a goal of creating a polyurethane circular economy. Chemical recycling creates a chemical mixture, requiring separation mechanisms to produce contaminant-free feedstocks. The proposed separation methods utilize membrane technology through a variety of processes such as pervaporation and filtration because these processes require lower energy. Pervaporation is a chemical potential driven process which uses dense polydimethylsiloxane (PDMS) membranes to remove any volatile components such as methanol, ethanol, butanediol, and ethylene glycol early in the process. Diffusion experiments show that all volatile components mentioned have diffusion coefficients on the same magnitude as ethanol, which is industrially separated through pervaporation. Pervaporation experiments provide permeability of each volatile component. Sorption coefficients represent the thermodynamic interactions as particles move through the membrane and are dependent on the diffusion coefficients and permeability. The proposed microfiltration process is pressure driven that uses porous bisphenol A polysulfone (BPA-PS) and 4,4′-dihydroxydiphenylmethane polysulfone (BPF-PS) membranes to remove larger components or oligomers such as polypropylene glycol (PPG). Microfiltration membrane process conditions are integral to the overall performance of membranes. Polymer composition, molecular weight, concentration, thickness, and operating pressures influence water permeance through the membrane. Higher molecular weight (~40 kDa) polymers at 15 wt% loading in N-methyl pyrrolidone (NMP) show the best performance for pure water.

This study investigates the separation of degradation products from both model and conventional polyurethanes. The model compounds considered are a poly(methylene diphenyl diisocyanate)-co-(1,4-butanediol) (pMDI-BDO) and a poly(MDI)-co-(bis(2-hydroxyethyl) terephthalate) (pMDI-BHET). The conventional polyurethanes consist of copolymers of either MDI or 2,4-toluene diisocyantate (TDI) with either poly(propylene glycol) (PPG) or poly(ethylene glycol) (PEG). Methanolysis of the MDI-BDO model compound yields a mixture of MDI-dimethyl carbamate (MDI-DC), BDO, a potassium tert-butoxide (KOtBu) base, methanol (MeOH), and water, from washing. Methanolysis of the MDI-BHET model compound can be controlled to give a mixture of MDI-DC, BHET, terephthalic acid (TPA), ethylene glycol (EG), KOtBu, MeOH, and water. Glycolysis of conventional polyurethanes gives a mixture of insoluble carbamate solids: MDI-DC or TDI-dimethyl carbamate (TDI-DC); water and MeOH soluble polyols: PPG or PEG; partially soluble diamines: 4,4-methylenedianiline (MDA) or 2,4-diaminetolune (DAT); and relatively volatile solvents and catalysts, 1,4-diazabicyclo[2.2.2]octane (DABCO), EG, MeOH, and water.

Leveraging the distinct physical and chemical properties of each component in these mixtures allows for a targeted separation strategy using the above membrane systems. Dimethyl carbamates insolubly in water and methanol allows them to be efficiently removed with microfiltration membranes. For the remaining components, ultrafiltration membranes are suited for capturing the slightly soluble molecules: the diamines and BHET. Ultrafiltration may also remove any soluble polyols. Liquid-liquid extraction is ideal for separating out the acids and bases: TPA and KOtBu. Pervaporation effectively targets the removal of the less volatile liquid components like BDO and EG. Finally, pervaporation can also be applied to the final separation of the water:MeOH mixture. To evaluate the effectiveness of microfiltration membranes, binary mixtures of each component with either MeOH or water were tested. Similarly, for pervaporation, binary mixtures of each liquid component (ex. BDO and EG) and MeOH or water were tested with PDMS membranes. Microfiltration effectively removes MDI-DC from both MeOH and water mixtures. BDO and EG permeate through microfiltration membranes, and are effectively captured by subsequent pervaporation, removing them from both MeOH and water.

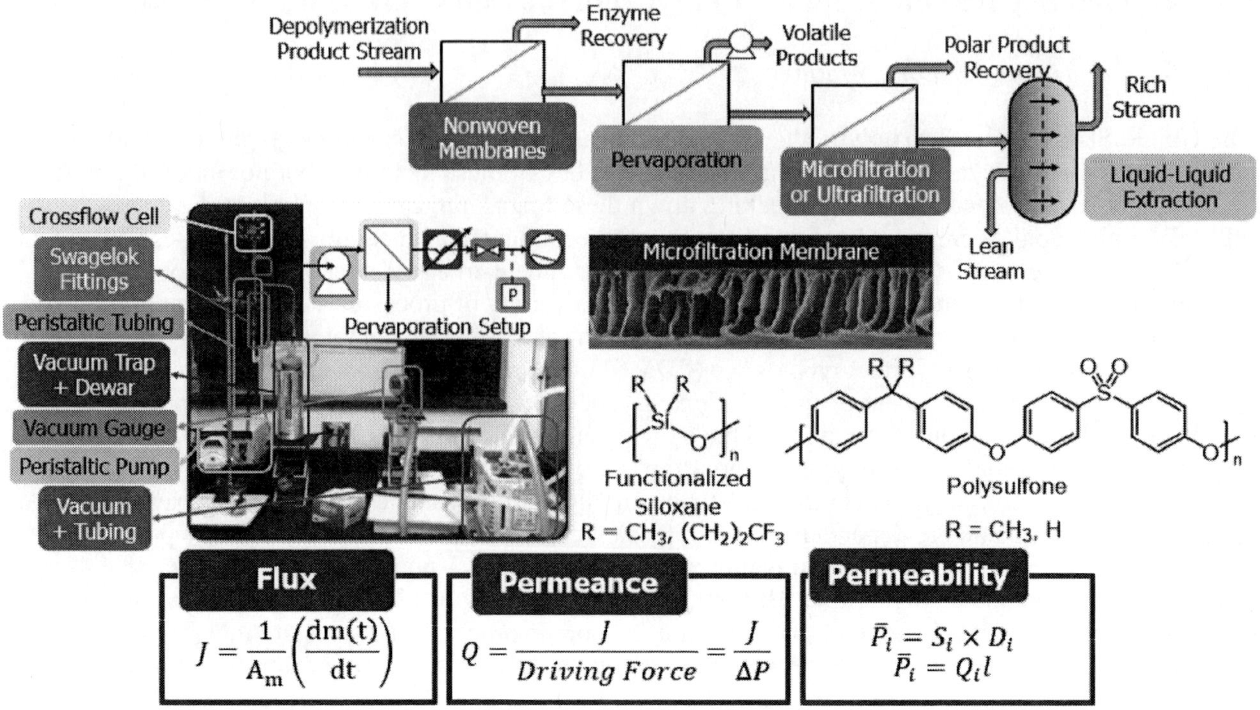
Depolymerization Product Stream
Enzyme Recovery
Volatile Products
Polar Product Recovery
Rich Stream
Nonwoven Membranes
Pervaporation
Microfiltration or Ultrafiltration
Liquid-Liquid Extraction
Lean Stream
Crossflow Cell
Swagelok Fittings
Peristaltic Tubing
Vacuum Trap + Dewar
Vacuum Gauge
Peristaltic Pump
Vacuum + Tubing
Pervaporation Setup
P
Microfiltration Membrane
Functionalized Siloxane
$R = CH_3, (CH_2)_2CF_3$
Polysulfone
$R = CH_3, H$
Flux
$J = \frac{1}{A_m}\left(\frac{dm(t)}{dt}\right)$
Permeance
$Q = \frac{J}{Driving\ Force} = \frac{J}{\Delta P}$
Permeability
$\bar{P}_i = S_i \times D_i$
$\bar{P}_i = Q_i l$

469a Evaluating Property Model Accuracy for Cost Optimization of Desalination Technologies

Savannah Sakhai, Timothy Bartholomew, Fernando Lima

Predicting water properties is an integral part of process-scale modeling and simulation for desalination technologies. However, since water chemistry is complex, property models can be intricate and pose significant computational challenges. Frequently, researchers use simplified property models to overcome these challenges, but the implications of these simplifications are not well known. This study aims to evaluate the performance of different property models to assess their impact on the efficacy of cost optimization models. The property models assessed include (i) a seawater property model, (ii) a sodium chloride property model, and (iii) an ideal sodium chloride property model. In this analysis, two prominent desalination technologies are considered: reverse osmosis (RO) and mechanical vapor compression (MVC), which encompass both membrane-based and evaporative desalination processes. Optimal design and operational variables are determined for treating seawater across a range of feed concentrations and water recoveries specific to each unit's application space. Results indicate that the sodium chloride property model yields comparable outcomes to the seawater property model across the operational domains for both RO and MVC processes. Conversely, the ideal property model exhibits significant discrepancies, although it performs better under isothermal conditions. Understanding the errors arising from specific modeling assumptions offers valuable guidance for decision-making and aiding the development of more sustainable and economically viable water treatment technologies.

469b Lithium Recovery from Mixed Brines Using Membrane Distillation-Crystallization

Olumide Ige, Sage Hiibel

The use of fossil fuels over the last century has altered the global ecosystem, causing health and environmental challenges as well as climate change. Due to increased concerns around these issues, coupled with the depletion of existing fossil fuel reserves, countries have adopted deadlines for transitioning away from fossil fuels to clean and renewable energy sources. This has led to the rise of electric vehicles and batteries, facilitating a large demand for lithium resources. Continental (salt lakes and salars), geothermal, and oilfield brines constitute about 66% of worldwide lithium sources, and these sources are typically more economical for direct lithium extraction than hard rock sources because they consume less water and produce a higher yield of lithium with fewer environmental impacts. However, the coexistence of different co-ions (i.e. K^+, Na^+, Mg^{2+}, and Ca^{2+}) with the lithium makes extracting high-purity lithium from these sources challenging due to similarities in chemical properties. Previous studies on the separation of lithium from other ions through solvent extraction and nanofiltration have notable drawbacks in terms of corrosion of equipment and membrane fouling. Membrane distillation crystallization is a promising hybrid technology that combines MD with a crystallization system to produce pure water (by MD) and high-purity crystals (via the crystallizer). MD is a thermally driven process water vapor transfers through a porous, hydrophobic membrane that prevents transfer of the liquid and non-volatile solutes. The objective of this study was to utilize the MDC system to selectively recover lithium from mixed salt brines. The effects of crystallizer temperature and water recovery on selectivity of the MDC process are also explored to maximize lithium recovery. The MDC system was shown to effectively removed potassium (99.6%) and sodium (99.06%) from the mixed brine. Along with research results, the principles, mechanisms, operations, process parameters, and industrial applications of MDC will be discussed in-depth in presentation.

469c Hydrothermal Liquefaction of Municipal Sludge: Inhibition of Heterocyclic N-Containing Organic Compounds from Aqueous Co-Product on Autotrophic Nitrifying Bacteria

João Poli, Adrian D. Romero, Susan Stagg-Williams, Sean Larson, Belinda S.M. Sturm

Waste-to-Energy (WtE) technologies hold significant importance in today's world, enabling the conversion of various types of waste into high-value products. Different feedstocks can be applied to a technology called Hydrothermal liquefaction (HTL). As the starting material undergoes HTL, it generates a mixture of solid, liquid, and gaseous products. Understanding the application of each HTL-resulting product is intrinsic and a highly investigated topic. The liquid fraction produced during an HTL reaction is called Aqueous-Co-Product (ACP), and an appropriate application has not been found yet due to its high organic composition and potential inhibitory compounds. Recycling the ACP back to the headworks of a Wastewater Treatment Plant (WWTP) is a potential effort, but understanding how that would affect the microbial community in terms of its inhibition potential remains an ongoing area of research interest. While previous research has indicated inhibitory effects of HNOCs on heterotrophic growth, their impact on autotrophs remains unclear. In this study, ACP produced from municipal sludge HTL has been characterized for its biological toxicity, considering the presence of heterocyclic N-containing organic compounds (HNOCs). This investigation delved into the inhibitory effects of HNOCs on autotrophic nitrifying bacteria by determining the half-maximum inhibitory concentration (IC_{50}) of ACP on nitrification. IC_{50} indicates the potency of a certain compound or substance in inhibiting a particular biological activity. Predominant HNOCs in ACP from municipal sludge were identified as 2-pyrrolidinone, 2-pyrazine, and 2-piperidinone and their derivatives, present at concentrations of 8.31, 8.22, and 0.96 mM, respectively. Individual testing revealed that 2-pyrrolidinone and 2-pyrazine hindered nitrifying bacteria by impeding ammonia uptake. Conversely, 2-piperidinone did not exhibit inhibitory effects on nitrification. Furthermore, testing of an ACP sample showed inhibitory effects on nitrification, with an IC_{50} of 0.08% (v/v), when the ACP was stored for 1 week after HTL reaction. Storage duration emerged as a significant variable, with ACP stored for 15 weeks showing reduced inhibitory effects on the nitrifying community. Specifically, the concentrations of HNOCs in the 1-week study decreased from 8.31, 8.22, and 0.96 mM to 6.36, 6.21, and 0.85 mM, respectively. The IC_{50} for the 15-week ACP was found to be 0.63% (v/v) which shows a clear reduction in its inhibitory effects on nitrifiers. In summary, this investigation underscores the inhibitory impact of ACP on nitrification, offering valuable insights for devising management strategies for the aqueous stream in municipal sludge HTL.

469d Integrating Graph Theory and Machine Learning to Design Reliable Wastewater Treatment Plants

Kirti Yenkie, Jean Pimentel, Emmanuel Aboagye, Akos Orosz, Heriberto Cabezas, Ferenc Friedler

Existing process synthesis and design approaches, such as superstructure optimization, are only as good as the user-defined technologies (process units), streams, and connections. The definition of these elements relies closely on heuristics, the experience of the design engineer, as well as reasonable initial guesses and bounds for the unknown variables [1], [2], [3]. It is expected that the final design resulting from these approaches will have adequate performance in terms of a predefined objective function. However, complex systems usually have multiple objectives and constraints that need to be satisfied while withstanding uncertainties. This creates additional difficulties in realistic problem formulation and existing solvers fail to converge to a global optimum. Thus, to demonstrate the utility of the integrated Graph Theory (GT) and Machine Learning (ML) approach, *'Design of Cost-effective, Sustainable, and Reliable Wastewater Treatment plants'* is selected as the problem of interest. Wherein, cost-effective implies minimum capital, operating, and maintenance costs [4]; sustainable implies maintaining ecological balance to support the well-being of current and future generations [5]; and reliable implies the minimum failure probability [6].

To this end, we have developed the Two-Layer Process Synthesis (TLPS) algorithm, which contains solvers for linear or nonlinear equations and efficiently handles the combinatorial part of the design problem. Thus, it will allow for timely convergence and optimal solution of complex and realistic process synthesis problems generating reliable optimal paths or flowsheets. The risk of failure of the associated infrastructure's elements is integrated into the TLPS algorithm through the application of machine learning classification algorithms on historical asset inventory, operating datasets, and weather forecasts to determine the failure probability. The ML-integrated TLPS algorithm is part of the enhanced process design framework, which will serve a two-fold purpose; (i) determination of structurally feasible, economically scalable, and sustainable paths for complex wastewater treatment (WWT) systems, and (ii) prediction of associated infrastructure reliability and resilience for proactive maintenance and management [7]. The TLPS resorts to the graph-theoretic approach known as the P-graph framework [8] for the identification of the n-best structures considering reliability and cost. Thus, a set of alternative designs is identified that provide insightful information about the system. The stakeholders will be able to redesign and retrofit existing facilities and plan maintenance schedules for safe and timely repairs with minimal investment. This will reduce operational downtime and lag between planning, adjustments, and implementation. Comparative assessments for WWT networks employing multiple modeling, optimization, and solution approaches will eliminate drawbacks and provide insights into non-intuitive solutions for enhancing process efficiency and maximum resilience.

References:

[1] H. Yeomans and I. E. Grossmann, "A systematic modeling framework of superstructure optimization in process synthesis," *Comput. Chem. Eng.*, vol. 23, no. 6, pp. 709–731, Jun. 1999, doi: 10.1016/S0098-1354(99)00003-4.

[2] J. D. Chea, A. L. Lehr, J. P. Stengel, M. J. Savelski, C. S. Slater, and K. M. Yenkie, "Evaluation of Solvent Recovery Options for Economic Feasibility through a Superstructure-Based Optimization Framework," *Ind. Eng. Chem. Res.*, vol. 59, no. 13, pp. 5931–5944, Apr. 2020, doi: 10.1021/acs.iecr.9b06725.

[3] W. Wu, K. M. Yenkie, and C. T. Maravelias, “A superstructure-based framework for bio-separation network synthesis,” *Comput. Chem. Eng.*, vol. 96, pp. 1–17, Jan. 2017, doi: 10.1016/j.compchemeng.2016.10.007.

[4] K. M. Yenkie, “Integrating the three E’s in wastewater treatment: efficient design, economic viability, and environmental sustainability,” *Curr. Opin. Chem. Eng.*, vol. 26, pp. 131–138, Dec. 2019, doi: 10.1016/j.coche.2019.09.002.

[5] U. Nations, “Sustainability,” United Nations. Accessed: Jul. 25, 2023. [Online]. Available: https://www.un.org/en/academic-impact/sustainability

[6] Z. Kovács, Á. Orosz, and F. Friedler, “Synthesis algorithms for the reliability analysis of processing systems,” *Cent. Eur. J. Oper. Res.*, vol. 27, pp. 573–595, 2019.

[7] Roshani E. and Filion Y. R., “Event-Based Approach to Optimize the Timing of Water Main Rehabilitation with Asset Management Strategies,” *J. Water Resour. Plan. Manag.*, vol. 140, no. 6, p. 04014004, Jun. 2014, doi: 10.1061/(ASCE)WR.1943-5452.0000392.

[8] F. Friedler, Á. Orosz, and J. Pimentel Losada, *P-graphs for Process Systems Engineering*. Cham, Switzerland: Springer International Publishing, 2022. doi: 10.1007/978-3-030-92216-0.

469e Bimetallic Nanoparticle Impregnated Surface Modified Activated Carbon for Application in Continuous Water Disinfection

Angira Das, Rajdip Bandyopadhyaya

The pressing global concern of providing safe drinking water is anticipated to worsen over the next decade due to the proliferation of pathogenic microorganisms and the emergence of antibiotic-resistant strains. Conventional centralized disinfection methods in urban areas, like chlorination, have drawbacks, such as generation of carcinogenic and mutagenic by-products. On the other hand, UV and membrane-based systems need significant capital requirements for operation; however, there is still a possibility of re-contamination during supply to the point of use (POU). Therefore, affordable POU water disinfection systems that can operate without electricity are urgently needed, especially in developing nations and regions afflicted by natural disasters where power infrastructure is often compromised. In this context, activated carbon (AC) finds widespread application in water disinfection systems, mainly attributed to its large surface area. With its significant adsorption capacity, AC effectively eliminates organic compounds, heavy metals, and a diverse range of contaminants that impart taste, colour, and odour to water; however, it can't eliminate bacteria.

Recent research shows impregnation of monometallic nanoparticles (silver, copper, gold, and titanium oxide) with well-established bactericidal properties onto AC for continuous water disinfection. More recently, combining monometallic particles to synthesize bimetallic counterparts has attained significant attention, offering tailored properties distinct from those of individual monometallic systems. This work explored bimetallic silver-copper nanoparticle (BMNP) impregnated on coconut shell-derived AC for possible superior bactericidal activity in a column-based system to facilitate continuous water disinfection.

Primarily, spherical BMNP were synthesized through a galvanic replacement reaction route using non-hazardous chemical reagents. Transmission electron microscopy (TEM) showed an unimodal size distribution of the particles with spherical morphology, and energy-dispersive X-ray spectroscopy (EDX) analysis verified a blended distribution of Ag and Cu atoms within individual particles. These nanoparticles were then impregnated onto the surface of AC with varying degrees of base functionalization. The generation of polar hydrophilic functional groups on base-treated AC was confirmed using FTIR spectroscopy, which facilitated the attachment of NP on AC. Successful impregnation of nanoparticles onto the AC surface was confirmed by observing characteristic FCC peaks in the X-ray diffraction (XRD) spectra for both Ag and Cu. The bimetallic NP-AC composite achieved a maximum loading of 3 wt.% Cu and 2 wt.% Ag, as measured using ICP-AES.

Batch mode experiments conducted in shake flask demonstrated that the composite was capable of 4 log reduction of *E. coli* cells in only 15 minutes of contact time, which was 2 and 1.6 times faster as compared to monometallic Cu-impregnated AC and monometallic Ag-impregnated AC, respectively. Furthermore, the increasing concentration of reactive oxygen species (ROS) over time verified bacterial cell death was attributed to the synergistic generation of ROS species by both Ag and Cu ions. The efficacy of the composite was further tested in a practical scenario within a continuous flow setup, packed inside a cylindrical column, typically as found in commercial household water purification systems. Studies revealed that the composite continuously eliminated 10^4 CFU/ml bacterial load in the simulated contaminated water successfully for at least 460 hours, yielding a minimum of 25 litres of potable water without attaining breakthrough. Simultaneously, the concentration of both Ag and Cu ions in the treated water remained within permissible limits, as prescribed by the USEPA and WHO. Thus, the developed bimetallic NP-impregnated AC composite exhibited promising antibacterial efficacy, providing an efficient solution for disinfecting water contaminated with *E. coli* and highlighting its potential to address water quality challenges in remote and disaster-affected areas.

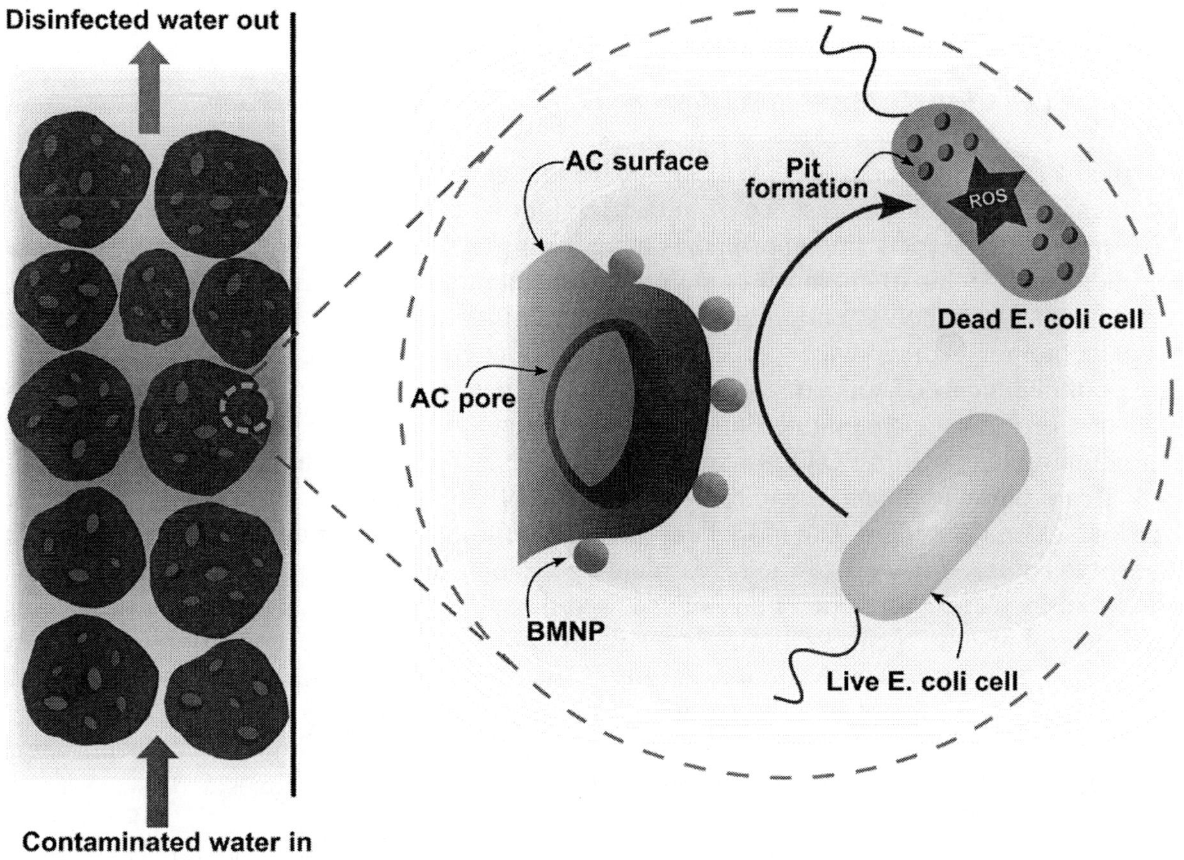

Schematic representation of bacterial inactivation in a unit of water inside a column packed with bimetallic nanoparticles impregnated activated carbon

469f Modeling and Estimating Chemical Releases from Industrial Wastewater Transfers to Publicly Owned Treatment Works

David Perez, John D. Chea, Gerardo Ruiz-Mercado, Tarek Abichou, Raymond Smith

Publicly Owned Treatment Works (POTWs) safeguard human and environmental health through wastewater management, yet the presence of industrial pollutants can drastically reduce the treatment efficiency. Several POTW modeling approaches have been developed to aid engineers and decision-makers in predicting the fate of chemical contaminants throughout the treatment process. However, these developments focus their attention on designing and modeling the wastewater organic content treatment and achieving effluent water quality criteria without considering potential air, land, and water releases of chemicals of concerns from industrial wastewater transfers. Therefore, this research effort proposes a process system engineering (PSE) approach to track and predict the fate of industrial wastewater contaminants within POTWs while supplying a standardized foundation that serves as a "benchmark" for predicting potential environmental releases, occupational exposure routes, and aid chemical risk assessment studies. The model captures interactions between chemicals of concern and individual process units across wastewater treatment operations, allowing for the estimation of chemical partitioning across unit operations such as sedimentation, biological treatment, adsorption, aerobic, and anaerobic equipment. Preliminary results for calculated chemical partitions to gas, liquid, and solid phases showed accurate predictions aligning with established benchmarks, literature studies, and operational data. The capability of this model to estimate contaminant partitions can provide risk evaluators with actionable information to calculate potential environmental and human health impacts and Environmental Justice (EJ) indexes affected by industrial wastewater releases.

469g Exploring Potential of Recycled Glass Sand As Water Filtration Medium

Shehbaz Ahmad, Fernanda Cedraz, Julie Albert, Tiong Aw

Each year, the United States dumps over 7.6 million tons of waste glass into landfills. At the same time, news headlines warn of sand shortages. Sand is not only the raw material for glass but is also used for construction, recreation, environmental restoration, and filtration, among other applications. Apart from recycling old glass into new glass, there are many other applications for crushed glass (or recycled glass sand) for which this vastly underutilized resource could be used, which would simultaneously reduce landfill waste and alleviate sand shortages. This study is evaluating recycled glass sand as a water filtration medium. This research involves a comprehensive examination of glass sand's permeability as a function of particle size distribution to elucidate its filtration capabilities for probe particles (<100 microns) and removal of *E. coli*. Laboratory scale filtration experiments were conducted using three different size sand particles. The particle sizes used for the experiment were L2 (3.4-1.7 mm), L3/L4 (1.7-0.4 mm), and L5-100 (0.4-0.1 mm). To attain a starting turbidity level of 75 NTU, 0.75 g of the probe material (glass sand particles < 100 microns) was suspended in 1000 mL of deionized water. As expected, we see a trade-off between turbidity reduction and permeability, with L5-100 providing the greatest turbidity reduction (3 NTU) but unacceptably low permeability ($9x10^{-10}$ m^2) and L2 providing an acceptable permeability ($2.8x10^{-8}$ m^2) but little turbidity reduction (50 NTU). This trade-off can be mitigated by layering the different glass sand sizes from largest (L2, top) to smallest (L5-100, bottom), resulting in turbidity reduction to 2 NTU and permeability of $4.8x10^{-9}$ m^2). For *E. Coli* treatment, natural water samples were spiked with *E. Coli to an initial concentration of approximately 400* colony-forming units per 100 mL (CFU/100mL). After filtration through the layered glass sand configuration, concentrations were less than 2 CFU/100 mL for all three replicates, representing log 2 removal (99.9% *E. coli* reduction). These findings show that recycled glass sand is a sustainable and effective filtration medium for contaminated water. Furthermore, the filter can be configured with different particle sizes and/or layers to meet application-specific requirements. This research is funded by the NSF Convergence Accelerator Program, Grant# 2230769.

469h Interactions of Graphene Oxide with the Microbial Community of Biologically Active Filters from a Water Treatment Plant

Tanvir Ahamed, Mengyan Li, Lisa Axe

With widespread occurrence and increasing concern of emerging contaminants in source water, biologically active filters (BAF) have been gaining acceptance in water treatment. Granular activated carbon (GAC), with its extensive surface area, micro-sized pores, and durable structure, provides a supporting substrate for advancing sustainable biofilms in BAF systems. In recent years, graphene oxide (GO) has been exploited to treat contaminants including atrazine, cephalexin, chloroquine, dipyrone, and drug-resistant bacteria and pathogens. Interestingly, in recent studies, GO has demonstrated biocompatibility through interactions with living organisms. Furthermore, with a surface area greater than GAC, the GO provides potentially more interactions between the substrate, contaminants, and microorganisms. As a result, GO in a GAC-BAF composite may enhance microbial growth and stimulate biodegradation in a BAF system. However, interactions between GO and the microbial community in water treatment processes such as BAFs remain unstudied. Therefore, in this research, we investigated the effect of GO on the properties and bacteria growth rate in a GAC-based BAF. The GO synthesized was characterized with a number of tools, including scanning electron microscopy (SEM), energy dispersive x-ray spectroscopy (EDX), X-ray diffraction (XRD), Fourier transform infrared spectroscopy (FTIR), and Raman spectrometry. Results demonstrated that GO exhibited the characteristic surface functional groups, particle size, and morphology. The GAC-BAF samples in this study were collected from the Passaic Valley Water Commission (PVWC) in New Jersey, that serves water to a significant portion of Northern New Jersey. Source water was prepared to simulate the influent water composition of the treatment plant and used as a culture medium for this study. The effect of GO on microbial growth was assessed through adenosine triphosphate (ATP) analysis by exposing the microbial community of GAC-BAF (200 mg/L) to GO concentrations (25 μg/L to 1000 μg/L) over 48 h. With an initial ATP concentration in the GAC-BAF of 26-70 ng ATP/cm^3, results showed an increase to 660 ng ATP/cm^3 after incubating in source water without GO (control). Interestingly, significant increases in ATP concentrations were observed by as much as 5% to 52% in the presence of GO compared to the control without GO. The growth rate was fit with the Monod equation revealing it exceeded the control by 20% to 46%. Specifically, the specific growth rate constant in the control batch $(0.111 \pm 0.01)\ h^{-1}$ increased by approximately 50% $(0.162 \pm 0.01)\ h^{-1}$ when exposed to 25 μg/L of GO. To address the potential toxicity of GO on the microbial community, reactive oxygen species (ROS) generation was measured using nitro blue tetrazolium (NBT) assay. During the exponential growth phase, compared to the control, ROS generation was not observed in the presence of GO. Further experiments are being conducted to characterize the BAF microbial community upon exposure to GO concentrations using Next-generation sequencing (Illumina MiSeq). This research provides new insights into the application of a bio-optimized BAF for advanced and sustainable water treatment or water reuse.

637a Microplastic Human Dietary Uptake from 1990 to 2018 Grew across 109 Major Developing and Industrialized Countries: Halving By Plastic Debris Removal

Xiang Zhao, Fengqi You

Microplastics (MPs) have been identified as emerging environmental and public health concerns over recent years [1]. These tiny particulates originate from the mismanaged plastic wastes from landfills and can be transported through air, water, and soil, posing environmental hazards to these earth counterparts and ecosystems. Industrial activities and developments can promote plastic waste generation and solid waste emissions to natural environments if mismanaged, exacerbating the MP level to surge by 50% in 2040 globally [2]. These tiny pollutants have been identified as potential public health hazards, and circumventing MP uptake demands holistic solutions to address both the existence and root causes of MP generation, which remains an unmet research need.

Navigating both the pathways and amounts of MP uptake from diet and air is pivotal for crafting effective strategies to mitigate human exposure and its potential health burdens [3, 4]. Dietary MPs involve those accumulated in foodstuffs and the material losses from plastic use in food and drink production, processing, and final product packaging [5], while airborne MPs mainly originate from the abrasion of plastic materials, such as those in tire and blown-ups from aquatic plastic particulates. One major source of aquatic MPs is mismanaged plastic waste runoffs from landfills or open dumping, which enter the surface water and generate macroplastic and MP via natural degradation [6]. These plastic particulates can contaminate water systems, but a holistic understanding of the aquatic MP amount transferred to humans remains uninvestigated due to the lack of explicit investigation into how these plastic particulates undergo nutrient transfer and are distributed at each trophic level. Dietary and airborne MP human uptake and their reduction strategies should consider local economies and industrial developments. As the UN Environmental Programme (UNEP) suggested, prevention of MP generation and exposure demands global collaboration in waste management, yet current practices mostly remain localized and lack a basic understanding of the spatial variability of plastic pollution [2].

Our work maps the MP transfer and exposure to humans by dietary and inhalation uptake in every five-year increment from 1990 to 2015 and in 2018 across 109 investigated countries by collating the existing investigation on MP exposure to humans with human air inhalation and domestic dietary habits. Our study found that rapidly industrializing countries, such as Indonesia, Malaysia, the Philippines, and Vietnam, topped the MP uptake globally, originating from high seafood consumption. These regions demand minimizing MP contents in raw foodstuffs, exposure through packaging materials, and release during food processing. Moreover, replacing single-use polyolefin plastic material with innovative, highly degradable alternatives, such as edible biobased materials, could help minimize the release of MPs from packaging materials, especially in beverages. Globally, African and Asian countries, such as Egypt, Indonesia, Malaysia, the Philippines, and Vietnam, identified as hotspots of MP uptake mainly from particulates in foodstuffs, demand government incentives to halt MP entry into the food web. Our study also captured the country-level MP exposure to provide technology implications for reducing its potential public health risks. Governments in developing and industrialized countries in Asia, Europe, Africa, and North and South America should incentivize the removal of free plastic debris from fresh- and saltwater environments through advanced water treatment and effective solid waste management practices.

References

[1] Elvis D. Okoffo, Erica Donner, Steve P. McGrath, Benjamin J. Tscharke, Jake W. O'Brien, Stacey O'Brien, Francisca Ribeiro, Stephen D. Burrows, Tania Toapanta, Cassandra Rauert, Saer Samanipour, Jochen F. Mueller, and Kevin V. Thomas. Plastics in biosolids from 1950 to 2016: A function of global plastic production and consumption. *Water Res*, 201:117367, 2021.

[2] Winnie W. Y. Lau, Yonathan Shiran, Richard M. Bailey, Ed Cook, Martin R. Stuchtey, Julia Koskella, Costas A. Velis, Linda Godfrey, Julien Boucher, Margaret B. Murphy, Richard C. Thompson, Emilia Jankowska, Arturo Castillo Castillo, Toby D. Pilditch, Ben Dixon, Laura Koerselman, Edward Kosior, Enzo Favoino, Jutta Gutberlet, Sarah Baulch, Meera E. Atreya, David Fischer, Kevin K. He, Milan M. Petit, U. Rashid Sumaila, Emily Neil, Mark V. Bernhofen, Keith Lawrence, James E. Palardy. Evaluating scenarios toward zero plastic pollution. *Science*, 369: 1455-1461, 2020.

[3] Wei Huang, Biao Song, Jie Liang, Qiuya Niu, Guangming Zeng, Maocai Shen, Jiaqin Deng, Yuan Luo, Xiaofeng Wen, and Yafei Zhang. Microplastics and associated contaminants in the aquatic environment: A review on their ecotoxicological effects, trophic transfer, and potential impacts to human health. *J Hazard Mater*, 405:124187, 2021.

[4] Qun Zhang, Elvis Genbo Xu, Jiana Li, Qiqing Chen, Liping Ma, Eddy Y. Zeng, and Huahong Shi. A review of microplastics in table salt, drinking water, and air: direct human exposure. *Environ Sci Technol*, 54(7):3740-3751, 2020.

[5] Gabriel Enrique De-la-Torre. Microplastics: an emerging threat to food security and human health. *J Food Sci Technol*, 57(5):1601-1608, 2020.

[6] Xiang Zhao and Fengqi You. Life cycle assessment of microplastics reveals their greater environmental hazards than mismanaged polymer waste losses. *Environ Sci Technol*, 56(16):11780-11797, 2022.

637b Elucidating the Role of Mechanical Forces in the Generation of Nanoplastics

Maninderjeet Singh, Nicholas Mendez, Michele Valsecchi, Sanat K. Kumar

Microplastics generation from plastic materials has emerged as one of the most challenging problems for human health in modern times. While microplastics pollution has been known for a few years now, the generation of nanoplastics and the adverse effects associated with nanoplastics have received less attention due to the difficulty of characterizing nanoplastics. However, nanoplastics are expected to cause significantly more health-related problems due to their ease of penetration to the body via ingestion and inhalation and their ability to enter cells and cross the blood-brain barrier, etc. Currently, the understanding of polymer physics pathways governing nanoplastics (and microplastics) generation and the structural architecture, and morphology of nanoplastics is not well understood. In this work, we elucidate the underlying mechanisms of nanoplastics generation from semi-crystalline and amorphous plastics under the effect of mechanical forces. We discuss the fragmentation pathways responsible for the generation of nanoplastics and their role in nanoplastics generation in bottled water, carbonated drinks, milk bottles, food packaging, etc. We discuss the possibility of the use of this knowledge to design plastics immune to the nanoplastics generation.

637c Environmentally Relevant Polystyrene Micro- and Nano- Plastics (MPs and NPs) Accumulation and Their Interaction with Lipids at Air-Sea Water (SW) Interface

Animesh Pan, Lauren Lamothe, Jeffy Wang, Natalie Paik, Geoffrey Bothun

Abstract

Despite increasing efforts to recycle plastic materials waste continue to accumulate in the oceans and it has been shown to accumulate in and negatively impact the health of marine life and added stress on ecosystems. Sea surface microlayers (SSMLs) is one of the critical natural components of oceans and a point of initial exposure to plastic pollution. SSML ranges from 1 to 1000 micrometer thick, are ubiquitous globally and are comprised of natural molecules, heavy metals, lipids, proteins, microorganisms, and particulate pollutants including microplastics (MPs) and nanoplastics (NPs). Persistent mechanical and photo-chemical degradation of plastics in the ocean yields micro- and nanoscale plastics (MPs and NPs) particles, which have been attracted much attention for their considerable hazards. It is equally important in determining how MPs and NPs are transformed within the SSMLs as transformations govern fate and transport. Detailed fundamental understanding of the SSMLs and impact of surface-active and particulate pollutants, are still lacking. In this study, we applied an accelerated weathering protocol to consumer plastics (Polystyrene (PS)) to create the engineered MPs and NPs using top-down method and investigated their colloidal properties in respect to their size, morphology, stability, and chemical composition under defined conditions. Engineered MPs and NPs have different average size fractions ranging from 5-10 μm, 1-5 μm, and < 1 μm with significant size and shape heterogeneity. We demonstrated the labeling approach of MPs with fluorescence dye (Nile Red) to facilitate the visualization and/or detection of MPs at the SSMLs. We also report their surface activity as well as their interaction with model lipid palmitic acid (PA) (abundant lipids in SSML) at the interface using Langmuir monolayer study. The intrinsic behavior of the MPs and NPs driving the surface activity and compressibility of the complexes at the air-lipid-SW interface. Our results provide physical and molecular characterization of a potentially significant MPs and NPs enrichment mechanism within the lipid monolayer at air-sea water interfaces.

637d A Microfluidic Platform for Enhanced Environmental Micro/Nanoplastic Analysis

Eric Johnston, Victor M. Ugaz

Degradation of polymeric materials into micro- and nano-scale particulates presents a daunting challenge for environmental monitoring and public health. The development of rapid, field-deployable detection methodologies would enable on-site analysis of pollutants, significantly impacting environmental monitoring and pollution mitigation strategies. However, traditional analytical techniques are constrained by their reliance on laboratory-based instrumentation, limiting portability and adaptability for field-based analysis. This presentation describes a microfluidic approach to characterize nanoplastic particles by size and concentration in a format with high portability potential. Our system employs a dual-stream microfluidic architecture where a tracer dye solution and a polymeric nanoparticle suspension are co-injected in parallel laminar flows. The interaction between these streams initiates a diffusion-driven complexation at their interface, yielding a unique fluorescence signature. Experimental validation using polystyrene nanoparticles in the challenging 20 – 100 nm size range demonstrates the system's capability to quantify particle size and concentration with high sensitivity. Finally, we introduce a physico-chemical model that enables the fluorescence signature to be predicted, making it possible to retrieve nanoparticle size, concentration, and composition in single- and multi-component mixtures. By allowing rapid micro/nano plastics detection, our work lays the foundation for a scalable solution to perform field-based environmental monitoring.

637f Computational Discovery of Plastic-Binding Peptides for Remediating Microplastic Pollution

Michael Bergman, Tianhong Tan, Fengqi You, Carol Hall

Methods are needed to remediate microplastic (MP) pollution so as to limit potential harm to the environment and human health. One promising approach is to use polypeptides to detect and/or capture MP pollution, as they can adsorb strongly to micro- and nanomaterial sized objects. Short peptides that bind to plastics are particularly interesting because their small size means that they are more economical to manufacture than proteins. However, peptides cannot currently be applied to MP remediation because there are no known peptides that bind to the many common plastics. In this work, we aim to fill this gap by using biophysical modeling and machine learning to discover 12-residue plastic-binding peptides (PBPs) that bind strongly to the common plastics polyethylene, polystyrene, polypropylene, and PET. We have been designed PBPs using the **Pep**tide **B**inding **D**esign (PepBD) algorithm. PepBD searches for PBPs by using Metropolis Monte Carlo sampling to randomly change the amino acid sequence or structure of a starting peptide adsorbed to a plastic surface. The changes are accepted or rejected based on the change in the peptide score, a measure of the peptide-plastic interaction energy and the stability (i.e., internal energy) of the peptide in the adsorbed conformation. Over the course of sampling millions of sequences and structures, PepBD identifies peptides with high affinity to the plastic. The best PBPs designs were found in molecular dynamics (MD) simulations to bind much more strongly to the target plastic than random amino acid sequences. Experimental testing supports our simulation results for the polyethylene design; designs for the other plastics are currently being evaluated. We next designed PBPs that have affinity for plastic as well as two other important features: high solubility in water, which facilitates MP remediation in aqueous environments; and preferential binding to a target plastic over others, which can help characterize or separate components of MP waste. The predicted solubilities were significantly improved by including the CamSol solubility score in the peptide score. The binding preference of a peptide for one plastic over others was moderately increased by training a long short-term memory (LSTM) network on PepBD data to predict the peptide's score given the amino acid sequence, and then searching for sequences with large score differences between a target plastic and an off-target plastic. By identifying PBPs for many common plastics, we hope that peptide-based methods for remediating MP pollution can begin to be developed.

637h A Novel Approach for Microplastic Removal from Aqueous Solution Via Fe_3O_4 functionalised g-C_3N_4 and Bnns: Recyclability and Toxicity Assessment

Riona Indhur, Arvind Kumar, Faizal Bux, Sheena Kumari

Introduction

The increase in MP concentration in water ecosystems continues to pose a serious threat to aquatic ecosystems, and subsequently human health as a result of bioaccumulation [1]. Significant number of MPs have been detected from different water matrices including surface water and wastewater treatment effluent, globally. Recent reports from South Africa have also indicated their presence in tap water in low concentrations [2]. Water treatment plants globally, are not optimised or designed for the removal of MPs [3]. A study by Kazour and team investigated the role of a municipal WWTP effluent as well as an abandoned coastal landfill as pathways for MPs insertion into the aquatic environment. They analysed the MPs in effluent, influent and sludge. They revealed that the WWTP exhibited a staggering daily discharge of ~ 227 million MPs [4]. In another study, Gies and teammates investigated the retention of MPs in a WWTP in Vancouver, Canada. They revealed that a staggering ~1.76 trillion MPs per year enter the urban WWTP. They investigated the plants efficiency at retaining these MPs and to decipher how much of MPs aren't retained and are released into the environment. Their study displayed that 99% of MPs that enter the WWTP had been retained. However, due to the extremely large volumes this major urban plant deals with, the 1% of MPs discharged into the environment equated to an astounding 30 billion MP particles being released into the sea [5]. Therefore, technological innovations are required to alleviate the limitations embedded in these treatment systems [6]. Nanotechnology has emerged as a pivotal technology to address a wide range of environmental challenges through performance enhancement [7]. Therefore, in the present investigation, microplastics were removed from treated wastewater and drinking water via nanomaterials g-C_3N_4-Fe_3O_4 and BNNS- Fe_3O_4, separately.

Methodology

The nanomaterials (g-C_3N_4-Fe_3O_4, BNNS- Fe_3O_4) were synthesized by thermal polycondensation method via calcination followed by the *in-situ* incorporation of Fe_3O_4 through conventional co-precipitation using ammonium hydroxide solution. Synthesized nanomaterials were further characterised by Zeta potential, XRD, TGA, FTIR, BET, XPS, SEM-EDX, TEM and VSM. Various parameters (pH, time, nanomaterial dose, microplastic dose) were optimised through gradient experiments. The microplastics utilised for microplastic dose consisted of a combination of PE and PS (PE=125 and 180 uM > PS > 25 uM). All batch experiments were performed in 500 mL flasks containing ultrapure deionized Millipore water (18.2 MΩ cm), appropriate nanomaterial dose and MP dose. All experiments were conducted at 25 °C. Optimised operating parameters were applied in batch experiments to investigate g-C_3N_4-Fe_3O_4 and BNNS- Fe_3O_4 for the removal of different types and sizes of microplastics (PE(125 uM), PS(180 uM > PS > 25 uM), PE + PS (PE=125 and 180 uM > PS > 25 uM), PS (500 uM > PS > 180 uM), PP(3 mm), LDPE(5 mm), HDPE(5 mm)) from Millipore water, wastewater effluent and drinking water. A recyclability study was performed wherein each nanomaterial underwent 5 successive rounds of reuse employing optimum operating parameters in 500 mL flasks. To assess the toxicity of the nanomaterials and wastewater effluent, phytotoxicity experiments were conducted to determine seed germination indices (G.I. %) utilising 3 different crops viz. *Vigna radiatus L.* (moong), *Cicer arietinum* (chickpea) and *Hordeum vulgare L.* (barley)

Results and Discussion

The maximum removal of PE (96.16%, size125 uM), PS (92.5% , size 180 uM > PS > 25 uM), PE + PS combo (94.89%, size PE=125 and 180 uM > PS > 25 uM), PS (45.62%, size 500 uM > PS > 180 uM), PP(0%, size 3 mm), LDPE(0%, size 5 mm) and HDPE (0%, size 5 mm) were noticed with catalyst g-C_3N_4-Fe_3O_4 under optimum operating conditions ([pH]= 4; time= 5 h; [nanomaterial]= 1.2 g/L; [MP]= 0.5 g/L) from UPW. The maximum removal efficiency of PE (94.44%, size 125 uM), PS(85.96%, size 180 uM > PS > 25 uM), PE + PS combo (88.28% , size PE=125 and 180 uM > PS > 25 uM), PS (38.77%, size 500 uM > PS > 180 uM), PP(0 %, size 3 mm), LDPE(0 %, size 5 mm) and HDPE(0 %, size 5 mm) were observed with catalyst BNNS- Fe_3O_4 under optimum operating conditions ([pH] = 3; time= 12 h; [nanomaterial]= 0.9 g/L; [MP] = 0.5 g/L) from UPW (Figure 1). A direct correlation can be made between the removal rate and the size of the MPs. This is evidenced by the results that exhibit a decrease in removal rate as MP size increases. Due to the MPs size, in the case of PP, LDPE and HDPE, the magnet was unable to remove the MPs from solution. The investigation of nanoadsorbent removal efficiency in different water matrices yielded 93.7 and 86.56 % from drinking water via g-C_3N_4-Fe_3O_4 and BNNS- Fe_3O_4, respectively. A removal rate of 91.91 and 83.78 % was observed from domestic wastewater effluent filtered with a 0.22 μM filter for g-C_3N_4-Fe_3O_4 and BNNS- Fe_3O_4, respectively, whilst a removal rate of 90.28 and 82.23 % was observed from the same domestic wastewater effluent (unfiltered) for g-C_3N_4-Fe_3O_4 and BNNS-Fe_3O_4, respectively (Figure 1). The results for filtered and unfiltered wastewater effluent are similar indicating that filtering plays no significant role in improving the removal efficiency. The decrease in removal efficiency between the drinking water and the wastewater effluent could be attributed to various ions present in the wastewater effluent. It is possible that the existing ions in the wastewater significantly affect the performance of the adsorbent on MPs.

The reusability study revealed that both nanomaterials retained a removal efficiency of more than 50% after 5 cycles whilst g-C_3N_4-Fe_3O_4 retained a removal efficiency of almost 80% (79.74 %) after 3 cycles. The phytotoxicity of domestic wastewater influent, effluent, g-C_3N_4-Fe_3O_4, BNNS- Fe_3O_4, g-C_3N_4-Fe_3O_4 filtrate and BNNS- Fe_3O_4 filtrate were assessed through seed germination indices (G.I.%). Direct discharge of g-C_3N_4-Fe_3O_4 and BNNS- Fe_3O_4 in solution are toxic, however, the nanomaterial filtrate of g-C_3N_4-Fe_3O_4 and BNNS- Fe_3O_4 revealed mild toxicity (approaching non-toxic) and no toxicity, respectively. This indicates that filtering the nanomaterial solution not only ensures recovery of the nanomaterial but greatly diminishes the environmental impact of these NMs. It also indicates that leaching of the nanomaterials into the filtrate is minimal.

Key conclusions

The present study demonstrates that both BNNS-Fe_3O_4 and g-C_3N_4-Fe_3O_4 are effective removers of MPs from wastewater and drinking water. The reusability study demonstrates that both nanomaterials are highly stable and maintained satisfactory results after the fifth cycle. The phytotoxicity study exhibits that although direct discharge of nanomaterials in solution is toxic, the nanomaterial filtrate of g-C_3N_4-Fe_3O_4 and BNNS-Fe_3O_4 are mildly toxic (almost non-toxic) and non-toxic, respectively.

References:

1. Dey, T. K.; Uddin, M.; Jamal, M., Detection and removal of microplastics in wastewater:

evolution and impact. *Environmental Science and Pollution Research* **2021,** *28* (14), 16925-16947.

2. Bouwman, H.; Minnaar, K.; Bezuidenhout, C.; Verster, C., Microplastics in freshwater environments–a scoping study. *Pretoria: Water Research Commission* **2018**, 3.
3. Iyare, P. U.; Ouki, S. K.; Bond, T. J. E. S. W. R.; Technology, Microplastics removal in wastewater treatment plants: a critical review. **2020,** *6* (10), 2664-2675.
4. Kazour, M.; Terki, S.; Rabhi, K.; Jemaa, S.; Khalaf, G.; Amara, R. J. M. P. B., Sources of microplastics pollution in the marine environment: Importance of wastewater treatment plant and coastal landfill. **2019,** *146*, 608-618.
5. Gies, E. A.; LeNoble, J. L.; Noël, M.; Etemadifar, A.; Bishay, F.; Hall, E. R.; Ross, P. S., Retention of microplastics in a major secondary wastewater treatment plant in Vancouver, Canada. *Marine Pollution Bulletin* **2018,** *133*, 553-561.
6. Hairom, N. H. H.; Soon, C. F.; Mohamed, R. M. S. R.; Morsin, M.; Zainal, N.; Nayan, N.; Zulkifli, C. Z.; Harun, N. H. J. E. T.; Innovation, A review of nanotechnological applications to detect and control surface water pollution. **2021,** *24,* 102032.
7. Goh, P.; Kang, H.; Ismail, A.; Khor, W.; Quen, L.; Higgins, D., Nanomaterials for microplastic remediation from aquatic environment: Why nano matters? *Chemosphere* **2022**, 134418.

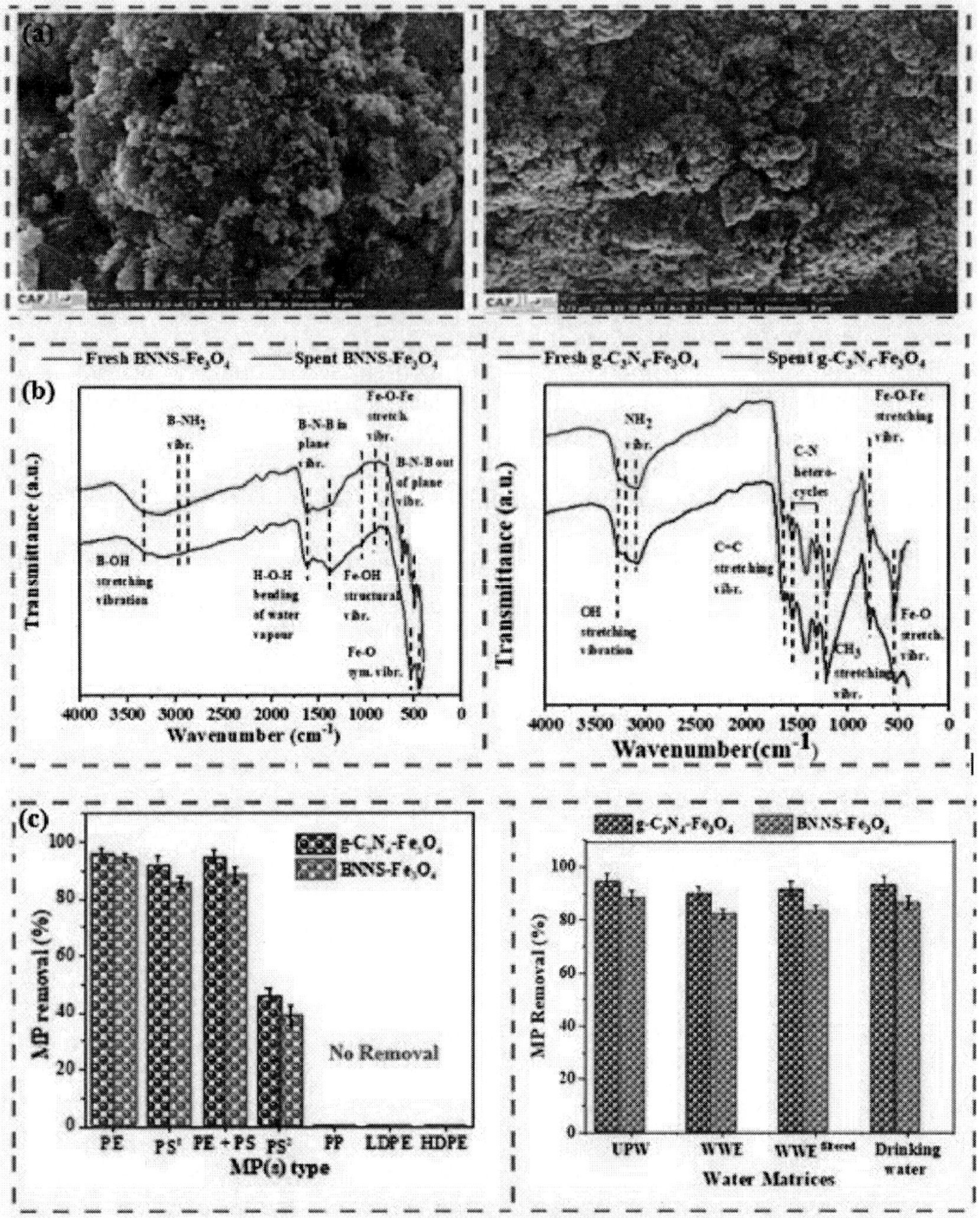

Fig. 1 (a) SEM of g-C_3N_4-Fe_3O_4 (left) and BNNS- Fe_3O_4 (right), (b) FTIR of fresh and used g-C_3N_4-Fe_3O_4 (right) and BNNS- Fe_3O_4 (left), (c) different size and type of MP removal from UPW (left) and MP removal from different water matrices.

321a Overview of per- and Poly-Fluoroalkyl Substances (PFAS) Research at the US Environmental Protection Agency

William Barrett

Per- and poly-fluorinated alkyl substances (PFAS) are a group of anthropogenic organic compounds that contain multiple CF_2 groups. The chemistry of the compounds makes them extremely useful in a wide range of chemical applications, and they have found use in a numerous consumer and industrial products. Another challenge with PFAS is that these CF_2 groups also make them problematic environmentally. These compounds are recalcitrant, difficult to break down, so they persistent in the environment. They impact human health and are ecological toxicants. Furthermore, the wide range of PFAS compounds complicates analytical methods to detect and characterize their presence in the environment. Lastly, recovery and destruction of these compound from the environment is difficult, making treatment of PFAS containing waste streams and remediation of contaminated sites difficult. This presentation will provide an overview of USEPA activities in the detection, characterization, treatment, and remediation of per- and poly-fluoroalkyl substances (PFAS).

321b Adsorption of PFAS Using Ion Exchanged Beta Zeolite

Charles Ponge, Nathaniel Sheehan, David R. Corbin, Edward Peltier, Justin Hutchison, Mark B. Shiflett

Per- and Polyfluoroalkyl substances(PFAS) widely produced since the 1940s have a broad range of uses in consumer and industrial products. The persistence of PFAS in the environment, and the associated health risks, have called for monitoring and remediation of PFAS in drinking water in the US and around the world. Zeolites have been shown to be effective sorbents. While traditional sorbent materials such as activated carbon and ion exchange resins are being used to remove PFAS from municipal drinking water, the materials are not easily regenerated in situ. Zeolites have been identified as sorbent materials with the potential to be regenerated on location and demonstrate selectivity for sorbing PFAS. Zeolites have a wide range of pore openings and tunable properties, making them ideal for removing both short- and long-chain PFAS from the environment. Currently, there is minimal research investigating zeolites for sorption of PFAS. The ion exchange of zeolites has been shown to change the properties of zeolites to increase catalytic or adsorptive capability. In this presentation, experimental data for ion exchanged zeolite CP814E will be shown and how the adsorption properties change with varying metals, and different loadings. The adsorption of PFAS is generally increased after ion exchange from the hydrogen form, with some metals showing 50-100% increase in adsorption in HPLC water. Thermal regeneration of these zeolites is also shown to recover most of the adsorptive capacity.

321c Understanding PFAS Selectivity in Anion Exchange Membrane Systems

Sam Thompson

The widespread production and use of multi-fluorinated carbon-based substances for a variety of purposes has contributed to the contamination of global water supply in recent decades. Per- and polyfluorinated alkyl substances (PFAS) are an anthropogenic class of chemicals imbued with desirable industrial properties that consequentially result in their difficult remediation once exposed to the environment. Common remediation strategies incorporate adsorption and/or ion exchange technologies to separate such contaminants from water, but often suffer from limitations due to low mass transfer efficiencies, among other issues. Membrane systems use many of the same separation methods currently in use today while enhancing operational potential of adsorption or ion exchange functionalities by boosting mass diffusion through convective flow. Furthermore, these improvements are applicable to the regeneration of such membrane systems, which is a costly drawback to conventional water purification techniques. To this end, commercial microfiltration and ultrafiltration membranes were modified using pore functionalization to incorporate an anion exchange moiety within the membrane matrix. This functionalization was performed with multiple amine-containing polymer networks ranging from weak to strong basic residues. Modified membranes maintained water permeance above 40 LMH/bar. Separation of PFAS from aqueous streams was analyzed for both "long-chain" and "short-chain" analytes, perfluorooctanoic acid and perfluorobutyric acid, respectively. Synthesized membranes demonstrated as high as 90% first-pass rejection of perfluorooctanoic acid. Ion exchange affinity of competitive solution-phase ions was also analyzed. This research advances a versatile membrane platform for environmentally relevant applications that seek to help the global availability of safe drinking water. Research funding was provided by the NIEHS-sponsored University of Kentucky Superfund Research Center.

321d Designing the Fluorinated Superfine-Activated Carbon Adsorptive Membrane for Selective Removal of Long Chain PFOS and Short-Chain Pfbs

Medha Kasula, Milad Esfahani

Membrane filtration and adsorption are two prominent methods for the effective removal of per- and poly-fluoroalkyl substances (PFAS) from water. Thin-film composite (TFC) nanofiltration (NF) membranes exhibit acceptable PFAS removal (<90%), but the presence of rejected PFAS in the concentrate stream necessitates an additional separation process. Our previous study reported the fabrication of adsorptive TFN NF membranes, incorporating superfine activated carbon (SFAC), for enhanced perfluorooctanesulfonic acid (PFOS) removal via adsorption during water filtration. The SFAC-TFN-0.5 membranes (0.5 wt% loading) of SFAC embedded inside the polyamide layer of TFN demonstrated 94% PFOS removal, with over 80% adsorption, and an increased permeability of 22 LMH/bar compared to the bare TFC. Although the results were impressive, the selectivity of the membranes towards different PFAS (long-chain PFOS and short-chain perfluorobutane sulfonate, PFBS), and in the presence of other components was not investigated. In this project, we investigated the selectivity of the adsorptive membrane toward different PFAS in the presence of NOM. Initial static adsorption results revealed limited PFBS adsorption on SFAC (45%), while it showed 93 % adsorption towards PFOS. To enhance the adsorbent's selectivity for different PFAS, SFAC was further fluorinated, as fluorination increases fluorophilic interactions with short-chain PFAS — an essential mechanism for their adsorption. The selectivity of PFOS in the presence of NOM on the SFAC membrane was observed to decrease to 70 % from 90 %. Therefore, the hypothesis is that fluorophilic interaction can boost the affinity of SFAC towards PFOS rather than NOM. In addition, the incorporation of SFAC in the IP process led to SFAC aggregation and disruption of the polyamide layer as reported in our previous study. In this project, fluorinated SFAC is grafted onto the active layer to prevent aggregation and better control surface properties.

321e Advancements in PFAS Separation and Concentration By Artisan Industriespresented By: Michael Fedorenko

Michael Fedorenko

Due to the physical nature of PFAS they are highly soluble and migrate easily in water resulting in a large amount of PFAS contaminated water that needs to be treated. With over 4,000 types of known PFAS, the chemistry constantly changing as new compounds are made, and the physical and chemical properties being largely unknown the remediation problem is difficult. Treatment levels for PFAS are extremely low, typically in the parts per trillion range requiring expensive activated carbon and/or ion exchange adsorbent treatment systems. PFAS remediation presents a significant challenge for low-income communities, especially as these treatment options become more expensive due to the cost of adsorbents and disposal increasing. Adsorbent remediation methods also generate a contaminated stream that needs to be dealt with. Historically, PFAS waste has been incinerated in the last decade and has further exacerbated the problem by not destroying the contaminants.

As manufacturers begin to phase out production of PFAS, due to their stability they will last in the environment for a long time and will continue to present remediation challenges. Technologies are being developed to destroy PFAS by breaking down the carbon fluorine (C-F) bond that makes them so stable. These technologies vary from supercritical H_2O or CO_2, advanced oxidation, plasma, catalyst, and others. The challenge for most of these techniques is equipment processing capacity. The flow rates needed by wastewater treatment plants and remediation sites are far greater than the highest capacity any destruction technology can achieve. Either the technologies need to be scaled up or the flow rates to the equipment need to be reduced.

Artisans has identified PFAS contaminants as a significant environmental problem lacking cost-effective treatment options and a final destruction solution to the forever chemicals. Artisan has been working for the last 3 years to develop a solution to remove all PFAS from aqueous sources. Then to find a way to incorporate PFAS destruction technologies into a continuous remediation train. Artisan believes that concentrating the PFAS stream is an effective solution to dealing with the high volumes of waste and will bridge the capacity gap for utilizing destruction techniques.

Artisan's presentation will cover:
1. Pilot Testing and Commercial Success – Demonstrate the effectiveness of the Artisan Rototherm®, a horizontal thin film evaporator that concentrates PFAS wastewater to a free-flowing powder and discharges clean water.
2. Recent Developments – Discuss Artisan process intensification via R&D and pilot trials on foam fractionation for PFAS concentration.
3. Future Work – Artisan Industries PFAS Destruction Train.

Artisan Industries is a technology and manufacturing company established and operated in Massachusetts, USA since 1934. Founded by MIT graduates, for over 90 years Artisan Associates have been at the forefront of solving difficult engineering challenges and manufacturing the solutions. Chemical engineers at Artisan rely on using proprietary process equipment, traditional unit operations, and developing new technologies to solve the problems at hand. Using Artisan's state-of-the-art pilot plant each process is tested and validated, data is generated for scale up design, and commercial systems are developed and built for professionals in many industries.

The presentation is given by Michael Fedorenko, Sr. Process Chemical Engineer at Artisan Industries. Michael has spent the last 10 years working on PFAS research, sampling, and commercial remediation. He received his B.S and M.S in chemical engineering at the University of Rhode Island College of Engineering. As a undergraduate and graduate student from 2014 – 2019 he researched PFAS physicochemical properties and worked as part of the STEEP group (Sources, Transport, Effects and Exposure of PFAS). He published an article titled: *Dominant entropic binding of perfluoroalkyl substances*

(PFASs) to albumin protein revealed by 19F NMR. After completing his M.S in 2019, he worked for a year and a half for Oneida ESG Group, in California on various PFAS environmental investigations and remediation projects at various military facilities. In 2021, Michael began working at Artisan Industries and has spent the last 3 years working on various R&D and commercial projects focused on PFAS remediation.

321f Chemically Modular Polymers for PFAS Remediation

Lakshay Dhamania, Joshua Moon, Rebecca Vosilla, Ethan Kermis

Per- and polyfluoroalkyl substances (PFAS) are synthetic chemicals that exhibit high thermal and chemical stability, rendering them valuable in various applications like non-stick coating, water-resistant clothes, cosmetics, paints, firefighting foam, etc. Unfortunately, these compounds are also known to bioaccumulate and exhibit long-term toxicity. PFAS are divided into major categories based on the structural differences of carbon chain and cationic head groups. Moreover, these compounds are known to transform into other PFAS compounds creating thousands of different PFAS. Two of the most ubiquitous PFAS are PFOA and PFOS that contain 8 carbon chain and a carboxylic and sulfonic head group, respectively.

In recent years, researchers have explored various methods for PFAS remediation, including activated carbon, ion-exchange resins, advanced oxidation processes, and filtration membranes. However, there remains a deficiency in adsorbents capable of effectively removing both long and short-chain PFAS compounds at environmentally relevant concentrations. While certain commercial sorption media exhibit efficiency in PFAS removal, they often suffer from drawbacks such as slow uptake kinetics and low sorption removal percentages for short chain PFAS.

To overcome these shortcomings of sorption media, we have developed chemically tunable polymer hydrogels that can effectively capture short and long-chain PFAS from aqueous media. These hydrogels are synthesized from poly (ethylene glycol) diacrylate and different monomer units of poly (ethylene glycol) methyl ether acrylate (PEGMEA), dodecyl acrylate, and pentafluoro phenyl acrylate (PFPA). Using active ester 'click chemistry', we can efficiently post-functionalize these polymers with Lewis base ligands by substituting PFPA groups in the networks. These ligands provide tunable acid-base interactions with anionic PFAS head groups. Quantification of the sorption efficiency and sorption isotherms has been performed using liquid chromatography-mass spectroscopy (LC-MS), with functionalized hydrogels showing as high as ~99% removal of PFOA from aqueous solutions. Sorption efficiency is further observed to be a function of the basicity of the grafted ligands. This chemically tunable approach will ultimately enable future structure-property studies for development of efficient polymer sorbents for PFAS remediation from surface and groundwater.

321g Development of Cationic Hydrogel PAC Composites for PFAS Remediation in Aqueous Systems and the G.I. Tract

Maria Victoria Ximenes Klaus, Molly Palmer, Kaitlyn Osborne, J. Zach Hilt

Per- and polyfluoroalkyl substances (PFAS) have been used in a wide variety of products and since resulted in accumulation in the environment and within our drinking water as a result of their persistent nature. Legacy PFAS compounds including perfluorooctanoic acid (PFOA) and perfluorooctanoic sulfonic acid (PFOS) have been connected to a myriad of harmful health impacts including decreased antibody response, dyslipidemia, and increased risks to several types of cancers. Current PFAS remediation treatments within the environment and our water sources include granular activated carbon (GAC) adsorption, separation by reverse osmosis and nanofiltration membrane systems, and formation of ionic bonds between anion exchange resin (AER) and negatively charged PFAS. Although these methods have high sorption capacity, some disadvantages including lack of selectivity and susceptibility to fouling which decreases efficiency, leading to the need for new cost-effective and efficient technologies for PFAS removal. Additionally, these techniques are limited to treating aqueous sources and not directly treating humans who have been exposed to PFAS. Enterosorbents are orally administered sorbents that bind to specific contaminants in the GI tract before excretion from the body. Novel materials based on synthetic sorbents, including hydrogels and hydrogel composites, have recently gained attention for remediation of these contaminants due to their high-water retention capacity and ease of tunability for increased PFAS affinity. Herein, cationic acrylamide-based hydrogels were synthesized using two cationic co-monomers, (3-acrylamidopropyl)ammonium chloride (DMAPAQ) and diallyldimethylammonium chloride (DADMAC), meanwhile, composites incorporated two particulates including powdered activated carbon (PAC) within the network. Materials were then characterized to confirm successful polymerization, and physicochemical analysis was completed, including Fourier-Transform infrared spectroscopy (FTIR) to discern its chemical composition, thermogravimetric analysis (TGA) to identify varying particulate loading percentage, and aqueous swelling for expansibility behavior. Sorbents were then subject to aqueous sorption studies to determine the affinity of PFOA and PFOS, and two short-chain analytes, perfluorobutanoic acid (PFBA) and perfluorobutanesulfonic acid (PFBS), to compare the impacts in sorption between long-chain and short-chain compounds, as well as measure influence of sulfonic versus carbonic functionality. FTIR confirmed expected functionalization of cationic materials, TGA analysis confirmed the varying concentrations of PAC and clay within the network, and the swelling ratio decreased with increasing crosslinking density and increased with the inclusion of DADMAC and DMAPAQ, as expected. Sorption of PFAS was detected via liquid chromatography tandem mass spectrometry (LM-MS/MS) where binding of PFAS was enhanced with increasing cationic functionality as well as with increasing particulate loading density. Finally, sorbent were analyzed in *in vivo* studies to determine enterosorbent efficacy with high safety and protection of species observed following PFAS exposure.

321h Advanced Adsorbents for PFAS Removal from Diverse Residual Streams

Ashley Butzlaff, Bineyam Mezgebe, Ashton Collins, Mohamed Ateia

Robust and selective treatment technologies, like adsorption, must be developed to effectively remove per- and polyfluoroalkyl substances (PFAS) from diverse matrices. Conventional adsorbents, such as granulated activated carbon (GAC) and ion exchange (IX) resin, have been shown to remove PFAS but performance can suffer in complex water matrices (e.g., salinity, organic content, inorganic ions). In this work, advanced adsorbents were investigated for residual stream PFAS management at bench-scale. Five advanced adsorbents – including functionalized cyclodextrin, polymer-metal oxide hybrid, and hydrogel – were evaluated for PFAS removal from residual streams. These advanced adsorbents are scalable and tunable (e.g., adjustable functional groups) with a small treatment footprint. Real residual water streams, including reverse osmosis (RO) concentrate and scrubber water blowdown, were selected to provide complex matrices to evaluate PFAS removal. The residual streams had diverse salinities and organic content, ranging from low conductivity ($< 500\ \mu S\ cm^{-1}$) and low organics (< 5 mg-C L^{-1}) to high conductivity ($> 2000\ \mu S\ cm^{-1}$) and high organics (> 80 mg-C L^{-1}), to holistically evaluate adsorbents for PFAS removal.

Kinetics, isotherm, and regeneration experiments were completed with the five adsorbents in the five residual streams. Multiple PFAS, with different chain lengths and functional groups, were investigated and quantified with targeted analysis. Conventional adsorbents, including granulated activated carbon (GAC) and ion exchange (IX) resin, were also evaluated in the residual streams to provide a performance benchmark. Results show that the advanced adsorbents provide fast and near complete PFAS removal across the diverse residual streams. Moreover, multiple regeneration cycles supported sustained performance and adsorbent reuse.

589a Exploring Molecular-Level Interactions in the Design of Redox Copolymers for Electrochemical Remediation of per- and Polyfluoroalkyl Substances (PFAS)

Anaira Roman Santiago, Song Yin, Johannes Elbert, Diwakar Shukla, Xiao Su

Exposure to anthropogenic contaminants has been a major driver for the development of more efficient technologies for water remediation. Recent advances point to electrochemical separations as a promising approach for water purification due to their modularity, scalability and potential for net-zero emissions when coupled with renewable energy sources. Despite these advantages, the implementation of electrochemical methods is often hindered by inadequate ion rejections, which increase the overall energy consumption. Thus, tailored materials are needed to provide molecular-level selectivity and achieve energy efficient separations. We have investigated molecular-level interactions to help drive the selective capture of per- and polyfluoroalkyl substances (PFAS). PFAS have emerged as an urgent target for water remediation due to their widespread prevalence and the enactment of new regulations by governmental agencies worldwide. Despite advancements in water treatment methodologies, the elimination of PFAS remains a formidable challenge due to their stable C-F bonds which confer amphiphilic characteristics and resistance to biological degradation, thereby necessitating energy-intensive conventional degradation approaches and specialized adsorbents.

In this study, we investigated the electrosorption of per- and polyfluoroalkyl substances (PFAS) utilizing functionalized redox-active copolymer electrodes incorporating 2,2,6,6-tetramethylpiperidinyloxy (TEMPO) and 2,2,3,3,4,4,4-heptafluorobutyl methacrylate (HFBMA) groups. The content of redox-active functional groups (TEMPO) was carefully balanced with fluorocarbon-containing groups (HFBMA) to elucidate the intricate effects of electrostatic and fluorophilic interactions. By combining results from electrosorption experiments with molecular dynamics (MD) simulations, we gained insights into the binding behavior of short-chain PFAS with different functional groups. Our findings demonstrate that fluorophilic groups can confer selectivity towards various PFAS depending on the length of the fluorinated chain of the target species. Experimental uptake capacities of PFAS were found to be consistent with MD simulations, highlighting the significance of electrostatic interactions for short-chain PFAS binding, with the strongest influence observed on chains with carbon atoms $C \geq 6$. Conversely, fluorophilic interactions were found to be favored by chains with $C \leq 5$. The electrodes exhibited notable removal efficiency of short-chain PFAS in secondary wastewater effluent (>70%), underscoring their practicality to remove PFAS under environmentally relevant conditions. Furthermore, we investigated the synergistic integration of electrosorption with electrooxidation as a feasible strategy for PFAS removal and destruction in wastewater, thus delineating a comprehensive approach towards sustainable PFAS remediation.

589b Removal and Recovery of Ammonium and Phosphate from Wastewater Using a Redox Flow Deionization Cell

Abdulrahman Alkhaldi, Zhenmeng Peng

Wastewater from domestic and industrial sources often contains significant amounts of ammonium (N) and phosphate (P). A high concentration of N and P can cause problems to the water body, leading to a series of issues like eutrophication and algal growth. On the other hand, N and P are valuable resources in many applications, with 90% of them being used as fertilizers for food production. With the world population increasing, these two nutrients are in huge demand to ensure food security. The synthesis of ammonia is energy intensive and causes one of the largest carbon emissions on a global scale. Phosphorous is a limited resource, with the reserves projected to deplete in 50–100 years at current usage rates. To make this worse, the phosphorous demand is expected to rise by 50–100% by 2050. Thus, removal and recovery of these nutrients from the wastewater would be both environmentally and economically beneficial.

Herein, we report the electrochemistry-based redox flow deionization cell (RFDC) to be utilized as a new technology for removal of recovery of N and P from wastewater. The study shows significant cell voltage and concentration effects on the cell performance. In both N and P removal experiments, an increase in the cell voltage leads to faster average salt removal rate (ASRR) and removal efficiency (RE) at cost of a higher electrical energy consumption (EEC), which can be attributed to excess energy consumption resultant of higher overpotentials to promote the redox kinetics. A cell voltage at about 1 V was found an appropriate value for balancing ASRR and EEC considerations. 4.64 $mmol_{NH4Cl}/m^2 \cdot min$, 99.1% and 105.3 $J/mmol_{NH4Cl}$ were achieved in the N removal experiment with an initial 10 mM NH_4Cl, and 2.81 $mmol_{Na3PO4}/m^2 \cdot min$, 80% and 273.7 $J/mmol_{Na3PO4}$ were achieved in the P removal experiment with an initial 7 mM Na_3PO_4 at 1 V cell voltage, demonstrating a fast, energy efficiency process for N and P removal. An increase in the initial N and P concentrations results in faster ASRR and lower EEC, which can be explained with a lower internal resistance that benefits with a lower Ohmic potential drop to drive the redox kinetics and a lower Ohmic energy loss and thus higher energy utilization efficiency. The effects of coexisting ions (Na^+ and Cl^-) on N and P removal and recovery were also investigated. The results show a relatively high N removal selectivity against Na^+ and a relatively low P removal selectivity against Cl^-, which can be attributed to a difference in the ion mass, size, and concentration parameters. The coexisting ions cause an increase in the EEC due to their simultaneous removal, which would be addressed by research of ion selective exchange membrane to suppress migration of coexisting ions. Nevertheless, regardless the presence of coexisting ions N and P can be removed all the way to the ppm level in the ions removal stream, well below the regulation limits, and in the meantime get recovered in the ions concentrated stream. Besides, the stability test confirms an excellent durability of the RFDC. All the findings suggest RFDC is effective and efficient for N and P removal and recovery from wastewater. With technological advantages of low cost, high efficiency, and continuous operation, this new technology has a great potential to be applied in domestic and industrial wastewater treatments to fulfill environmental and economic goals.

589c Advancing the Efficiency of Plasma-Based Water Treatment: Insights into Ion Effects on Contaminant Degradation

Foluke Ganzallo, Selma Mededovic

Plasma-based water treatment (PWT) has emerged as a promising technology for the degradation of organic and inorganic contaminants in various water sources. In PWT, an electrical discharge generates a wide range of reactive species by exciting and splitting water and background gas, enabling the in-situ formation of oxidative and reductive species without the need for chemical additives. This feature makes PWT an attractive and superior option for numerous water treatment applications. However, the performance of PWT reactors when applied to treat groundwater, wastewater, membrane concentrate, or ion exchange brine is influenced by the physicochemical properties of the solutions, with particular emphasis on solution electrical conductivity. Recent investigations into treatment of short- and long-chain poly- and perfluoroalkyl substances (PFAS) in ion exchange regenerant still bottom solutions and membrane concentrates, characterized by high solution electrical conductivities, have revealed that ion concentration plays a pivotal role in determining treatment effectiveness. This influence arises from its impact on the physical and chemical processes occurring at the plasma-liquid interface. This study explores the roles that inorganic ions play in bulk liquid and plasma chemistry by closely examining the transformation pathways of various alkali metal halide and ammonium halide salts. The insights gained from this research are extended to the treatment of industrial PFAS matrices in solutions characterized by high electrical conductivity.

589d Using Data-Driven Modeling and Systems Optimization for Advancing Sustainable Nutrient Recovery Technologies from Concentrated Wastewater Sources

Md Masud, Thisali Wadunambu Arachchige Dona, Km Prottoy Shariar Piash, Oishi Sanyal, Yuhe Tian

The discharge of nitrogen (N), phosphorus (P), and potassium (K) from wastewater treatment plants contributes significantly to nutrient pollution, leading to harmful algal blooms and eutrophication in freshwater ecosystems [1,2]. On the other hand, as per recent statistics [3], nearly 13% of the global nutrient demand can be supplemented via the recovery of these key fertilizer nutrients from wastewater. Nutrient recovery from wastewater thus offers a win-win solution of food and environment. Toward this direction, a novel membrane separation process has been experimentally developed in the team's previously work [4], demonstrating superior N/P/K separation efficiency and energy savings. In this presentation, we aim to quantitatively model and analyze this process to identify the optimal process design conditions.

Machine learning-aided modeling is first leveraged to unravel the structure-property-process relationships governing membrane performance based on experimental data encompassing various membrane properties, process conditions, and performance metrics. To address the challenge of limited data from experiment, the Synthetic Minority Oversampling Technique (SMOTE) is utilized to generate synthetic data based on the original experimental data [5]. A neural network (NN) model is then trained to extract underlying nonlinear input-output relationships. This SMOTE-integrated NN modeling approach has been demonstrated in a previous study for microwave-assisted ammonia production, and we aim to extend its application to membrane separations [6]. The resulting data-driven structure-property-process relationships are further integrated with first-principles membrane models, from which sizing and costing can be evaluated. On this basis, process design optimization is conducted to assess the economic viability of our proposed nutrient recovery process at an industrial scale. Inverse design is also performed to identify the corresponding structural parameters for membrane synthesis. Our analyses strive to provide insights into process bottlenecks or opportunities to enhance economic competitivity and production scalability of such sustainable nutrient recovery processes.

Reference:

[1] Chislock, M. F., Doster, E., Zitomer, R. A., & Wilson, A. E. (2013). Eutrophication: causes, consequences, and controls in aquatic ecosystems.*Nature Education Knowledge*, *4*(4), 10.

[2] Preisner, M., Neverova-Dziopak, E., & Kowalewski, Z. (2021). Mitigation of eutrophication caused by wastewater discharge: A simulation-based approach.*Ambio*, *50*(2), 413-424.

[3] Facts about wastewater and nutrient management. United Nations Environmental Programme.

[4] Piash, K. S., Anwar, R., Shingleton, C., Erwin, R., Lin, L. S., & Sanyal, O. (2022). Integrating chemical precipitation and membrane separation for phosphorus and ammonia recovery from anaerobic digestate. AIChE Journal, 68(12), e17869.

[5] Chawla, N. V., Bowyer, K. W., Hall, L. O., & Kegelmeyer, W. P. (2002). SMOTE: synthetic minority over-sampling technique. *Journal of artificial intelligence research*, *16*, 321-357

[6] Masud, M. A. A.; Araia, A.; Wang, Y.; Hu, J.; Tian, Y., Machine Learning-Aided Process Design for Microwave-Assisted Ammonia Production. Computer-Aided Chemical Engineering. Under review

589e Synthesis and Characterization of Fluorinated Magnetic Hydrogel Nanocomposites for PFAS Remediation in Aqueous Systems

Pranto Paul, J. Zach Hilt

Decades of use of per- and polyfluoroalkyl substances (PFAS) in a multitude of consumer and industry-based products have led to a devastating amount of soil and water contamination. The chemical and thermal stability of PFAS have proved to be an especially daunting challenge from an environmental remediation standpoint. Presently, the only full-scale water treatment separates via sorption and uses non-selective materials such as activated carbon (AC) or mineral media which are extremely difficult and/or costly to regenerate. Developing effective and renewable remediation technologies that lead to PFAS free water sources are, therefore, a vital part of current research efforts. Research focused on selective sorption is becoming a more practical route for capture and removal from contaminated water systems. Novel materials based synthetic sorbents such as hydrogel composites have recently gained attention for remediation of these contaminants due to their ability to be functionalized to target PFAS sorption. Additionally, magnetic nanoparticles (MNPs) embedded in the hydrogel matrix named as magnetic hydrogel nanocomposite enable energy to be delivered remotely to the material using an alternating magnetic field (AMF). This localized energy can be used to drive processes to enhance material performance, including disruption of binding interactions. This combination of functionalized hydrogel and MNPs addition could be a sustainable solution for PFAS removal via sorption and desorption. Herein, perfluoro monomer, 2,2,2-Trifluoroethyl acrylate (TFEA), based MNP embedded hydrogel nanocomposites were synthesized with the variation of TFEA and MNPs percentage in the composite. Physicochemical characterization, including thermogravimetric analysis (TGA), and AMF heating were performed and confirmed MNP loading, and aqueous swelling was performed and confirmed TFEA addition. Finally, aqueous sorption studies of PFAS were performed to determine the affinity of two long chain PFAS compounds, perfluorooctanoic acid (PFOA) and perfluorooctanoic sulfonic acid (PFOS), and two short-chain PFAS, perfluorobutanoic acid (PFBA) and perfluorobutanesulfonic acid (PFBS). Significant binding of long chain PFAS was found and promising sorption of short chain PFAS were found.

589f Efficiency and Feasibility Assessment of Calcined LDH Pyroaurite for Phosphorus Removal from Municipal Wastewaters: Experimental and Simulation-Based Investigation

Carla Maggetti, Benedetta Martellotti, Giacomo Antonioni, Valerio Cozzani, Tommaso Tabanelli, Fabrizio Cavani, Davide Pinelli, Dario Frascari

Introduction

Phosphorus (P) has been identified as a critical raw material by the European Union due to its crucial role in vegetation growth [1]. However, excessive P presence can lead to harmful phenomena such as eutrophication. To address this, the EU has introduced threshold values, 1 mg L^{-1}, for P concentrations in municipal wastewater (MWW). Balancing agricultural needs with sustainable environmental practices is challenging, with MWW emerging as a potential solution containing about 7 mg_P L^{-1}. Current MWW treatment methods for P removal have drawbacks. Chemical precipitation requires expensive reagents and may generate waste, while biological processes are sensitive to environmental conditions and advanced oxidation processes can be energy-intensive. In this context, the adsorption and ion exchange process presents advantages for P removal and recovery. It effectively removes P even at low concentrations, operates in various environmental conditions, and allows for regeneration and reuse of adsorbed P. Additionally, it doesn't require the addition of chemical reagents, reducing the risk of harmful by-products. Furthermore, nutrient removal and recovery via ion exchange have proven economically advantageous compared to traditional biological nutrients removal, which predominates in MWW treatment plants (MWWTPs), with a total cost being lower [2]. Various materials have been experimented with for P removal from MWW, among which Layered Double Hydroxides (LDHs) stand out due to their high adsorption capacity and ability to be regenerated for continued use. LDHs contribute to reducing operational costs and environmental impact in MWWTPs.

Therefore, this study investigates the effectiveness of calcined LDH pyroaurite for P removal from MWWs. It pioneers continuous-flow ion exchange processes tailored for P removal and recovery in real MWW matrices by exploring the adsorption and ion exchange properties of calcined LDH pyroaurite, evaluating its mechanical resistance and chemical stability through repeated cycles of adsorption and desorption, and assessing the performance of the regeneration and P recovery process from the adsorbent material. Additionally, this study aims to scale up the process from a laboratory-scale plant to a pilot-scale plant.

Furthermore, all the experimental data collected were used to simulate the process using the Aspen Adsorption™ v14 software. Aspen Adsorption™ accurately simulates solid-liquid phases, optimizing adsorption processes. Its multifunctionality includes advanced equations of state and thermodynamic algorithms, facilitating exploration of various process parameters. Despite limited studies on ion exchange, its potential for efficiency enhancement and sustainability in industrial operations is evident, allowing for a reduction of costs and fostering innovation.

Methods

The main characteristics of LDH pyroaurite are described in [3]. The experimental procedures and methods employed are detailed in [4]. Continuous tests were conducted in a laboratory-scale PVC column measuring 0.013 m in diameter, with the adsorbent material packed to a height of 0.20 m. At the conclusion of preliminary bench-scale testing, the material underwent pilot-plant scale testing in a fully automated facility using a column with a diameter of 0.013 m and a bed height of 0.60 m. All tests were conducted using real MWWs obtained from a MWWTP based in Northern Italy. In the Aspen Adsorption™ software, input parameters inserted are listed in Table 1 and 2.

Results

Isotherm tests were conducted using real MWWTP effluent as the liquid medium, with an adsorbent concentration of 1 g L^{-1}. The material was tested in its original state and after repeated BT tests (Figure 1), with isotherms fitted using Langmuir and Freundlich models. The Langmuir model provided the most accurate fit, with model parameters detailed in Table 3. Experimental data indicated favorable isotherms, with high solute adsorption even at low concentrations. The maximum sorption capacity at equilibrium (C_s^{∞}) was estimated at 20.7 mgP $g_{dry\ adsorbent}^{-1}$ for the virgin material. Differences between isotherms before and after BT tests were minimal and not statistically significant, demonstrating pyroaurite sustained adsorption capacities over multiple cycles. Compared to a hybrid anion exchanger Layne^{RT} [5], here showed as a benchmark, calcined pyroaurite exhibited superior C_s^{∞} and K_{eq} values, resulting in a steeper and more favorable isotherm.

BT tests for adsorption/desorption were conducted using a packed-bed column fed with effluent. During the adsorption phase, an empty bed contact time (EBCT) of 5 minutes was employed. The results depicted in Figure 2a exhibit remarkable selectivity towards P: all other anions are eluted prior to the appearance of phosphate. Competitive anions exit the column between 8 and 12 hours of operation, returning to inlet concentrations by hour 29. P, on the other hand, begin to elute after approximately 35 hours, indicating the potential for treating substantial MWW within a single adsorption cycle. Pyroaurite demonstrated very good stability, retaining its performances over 4 repeated BT tests, achieving a P adsorption yield of 97% at the breakpoint (BP) 1 mg_P L^{-1}, with a sorbent utilization efficiency at BP of 67% (Table 4).

The desorption/regeneration procedure was performed eluting the NaOH 0.5 M with EBCT 10 min to regenerate the resin and recover the P rich product (Figure 2b). Phosphates and sulphates co-eluted, with phosphate concentrations reaching about 690 mg L^{-1}. Chloride desorbed at much lower concentrations, while nitrate levels were minimal. By cutting the P-rich fraction at 28.5 hours, a solution with 210 mg L^{-1} of P and 237 of SO_4^{2-} concentration is obtained, achieving a 34-fold concentration of P, but only 2.4 of SO_4^{2-}. The P desorption yield resulted to be 65%.

Figure 2a illustrates a comparison between the P adsorption BT curves achieved with bed heights of 20 and 60 cm. The elution sequence remains consistent, with comparable separations between the BT curves of various anions (not depicted). Notably, a slight delay in BT was observed for P during the adsorption phase. A comparison between the performance parameters obtained from the 2 BT tests are shown in Table 4. These results offer promising insights into the feasibility of scaling up the P adsorption process.

The data collected from isotherm and BT tests conducted were employed to perform the simulation of the process with Aspen Asdorption^{TM} software. The model employed for simulating the BT curve was the Ion Exchange model, which also allows for the simulation of adsorbent regeneration. The choices made to configurate the bed (Table 1) determined the set of equations governing the simulation of the adsorption process. The equation of the isotherm considered, along with its explanation, is provided in Table 1, and the associated parameters inserted as inputs were: the valence ratio between A and B (m) for all ions and the total capacity of the material (Q) (Table 2). Specifically, the value of the Q was obtained experimentally. The equilibrium constants of the ionic compounds (K_{AB}), the Mass Transfer Coefficient (MCT) and the initial concentration of bicarbonate (HCO_3^{2-}, not determined experimentally) were estimated using the dedicated function of the software (Table 5). As evident from Figure 2a, a good best-fit was obtained for all ions except for NO_3^-. Indeed, due to its very low concentrations in the MWW, NO_3^- was not included in the objective function that was minimized. The correspondence between simulated and experimental curves confirms the sound choice of the model and input parameters, thereby enabling further parameter adjustments to optimize the process in terms of performance and costs prior to conducting additional experimental tests.

Conclusion

The calcined LDH pyroaurite demonstrated excellent performance in adsorbing P from MWW effluent, as evidenced by favorable isotherms, extreme selectivity for P, and sustained adsorption capacities over multiple cycles. Compared to HAIX Layne^{RT}, calcined pyroaurite exhibited superior sorption capacity and equilibrium constant values, indicating a more favorable isotherm. The adsorption/desorption BT tests highlighted pyroaurite's remarkable selectivity towards P, with efficient removal of competing anions. Furthermore, the desorption/regeneration procedure successfully recovered P-rich products, achieving a significant concentration increase. These findings suggest the potential for scaling up the P adsorption process using pyroaurite, promising advancements in MWW treatment technologies.

Aspen Adsorption^{TM}, employing the Ion Exchange model, demonstrated excellent simulation of adsorption processes, guided by experimental data for precise configuration. Despite challenges fitting NO_3^-, agreement with other ions affirmed model efficacy. Estimating equilibrium constants from BT curves enhanced predictability, showcasing Aspen Adsorption^{TM} 's versatility in system optimization. Next, focus will shift to the simulation of the material regeneration step and process scale-up simulation. Once a robust parameter set is established, optimization aims to boost overall efficiency. This iterative approach holds promise for refining adsorption system design for practical applications.

References

[1] «Critical raw materials - European Commission».

[2] X. Huang, *et al.*, «Economic evaluation of ion-exchange processes for nutrient removal and recovery from municipal wastewater», *Npj Clean Water*, vol. 3, fasc. 1, pp. 1–10, 2020, doi: 10.1038/s41545-020-0054-x.

[3] F. Cavani, *et al.*, «Hydrotalcite-type anionic clays: Preparation, properties and applications.», *Catal. Today*, vol. 11, fasc. 2, pp. 173–301, 1991, doi.org/10.1016/0920-5861(91)80068-K.

[4] D. Pinelli *et al.*, «Ammonium recovery from municipal wastewater by ion exchange: Development and application of a procedure for sorbent selection», *J. Environ. Chem. Eng.*, vol. 10, fasc. 6, p. 108829, 2022, doi: 10.1016/j.jece.2022.108829.

[5] D. Pinelli *et al.*, «Regeneration and modelling of a phosphorous removal and recovery hybrid ion exchange resin after long term operation with municipal wastewater», *Chemosphere*, vol. 286, p. 131581, 2022, doi: 10.1016/j.chemosphere.2021.131581.

Table 1. Aspen Adsorption™ bed configuration.

General	Discretization Method To Be Used	UDS 1
	Number Of Nodes	20
Material/Momentum Balance	Material Balance Assumption	Convection with Estimated Dispersion
Kinetic Model	Film Model Assumption	Fluid
	Kinetic Model Assumption	Lumped Resistance
	Form Of Lumped Resistance Model	Linear
	Form Of Mass Transfer Coefficient	Constant
Isotherm	Isotherm Assumed For Layer	Mass Action Equilibrium

Table 2. Isotherm inputs parameters in Aspen Adsorption™

m PO4 (-)	2[a]
m SO4 (-)	2
m HCO3 (-)	1
m Cl (-)	1
m NO3 (-)	1
Q (eq/m³)	4300

[a] The phosphate ion has been assigned a valency of +2 because at the working pH range of 7.5-8.5, the phosphate ion exists in a mixture of HPO_4^{2-} and $H_2PO_4^-$, with a distribution skewed towards HPO_4^{2-}.

The exchange reaction in the ion-exchange process typically takes the form:

$A + mBR \Leftrightarrow ARm + mB$

Where :

- m is an integer or a fraction. It is given by the valence ratio of A and B
- A refers to an ionic component in solution
- B refers to a counter-ion on the ion-exchanger surface
- R refers to a bound group (of opposite sign to B)

The associated equilibrium relationship can be written as:

$$\frac{1}{K_{AB}} = \left(\frac{y_A}{x_A}\right)\left(\frac{x_B}{y_B}\right)^m\left(\frac{Q}{c_0}\right)^{m-1}$$

K_{AB} = Equilibrium constant or selectivity coefficient

x = Equivalent mole fraction in the adsorbed phase

y = Equivalent mole fraction in the aqueous phase

c_0 = Total ionic concentration

Q = Ion-exchange resin capacity

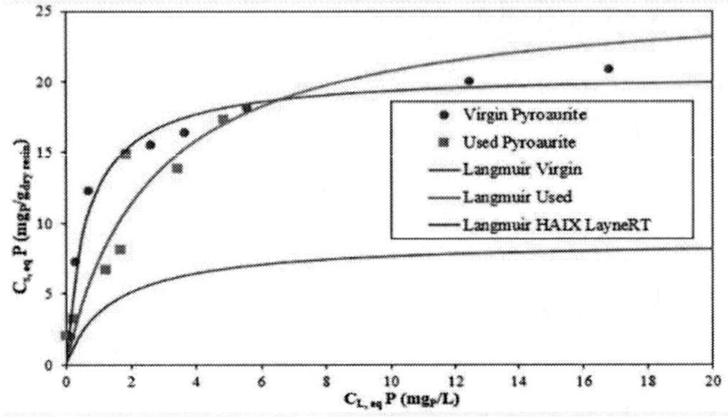

Table 3. Isotherm fitting with the Langmuir and Freundlich models. Best-fitting model parameters and P adsorption capacity in equilibrium with a liquid phase concentration of 7 mgP L-1.

		Units	Vyrgin Pyroaurite	Used Pyroaurite	Virgin HAIX Layne[RT]
Langmuir model	Cs^{∞}	$mg_P\ g_{dry\ sorbent}^{-1}$	20,65	26,27	8,70
	K_{eq}	$L\ g^{-1}$	1,50	0,38	0,70
	R^2		0,97	0,86	0,88
Freundlich model	n		3,86	1,91	
	K_F	$mg^{(1-\frac{1}{n})}\ g^{-1}\ L^{(\frac{1}{n})}$	10,78	7,58	
	R^2		0,88	0,87	

Figure 1. Isotherms conducted with the Northen Italy MWW, using LDH Pyroaurite in its virgin (blue circles) and used (red squares) conditions. Continuous lines correspond to the best-fitting simulation performed with the Langmuir model. Langmuir model simulation curve for HAIX LayneRT resin is shown (green line) for comparison. CL,eq: liquid-phase P concentration at equilibrium (mgP L-1); CS,eq: solid-phase P concentration at equilibrium (mgP gdry resin-1).

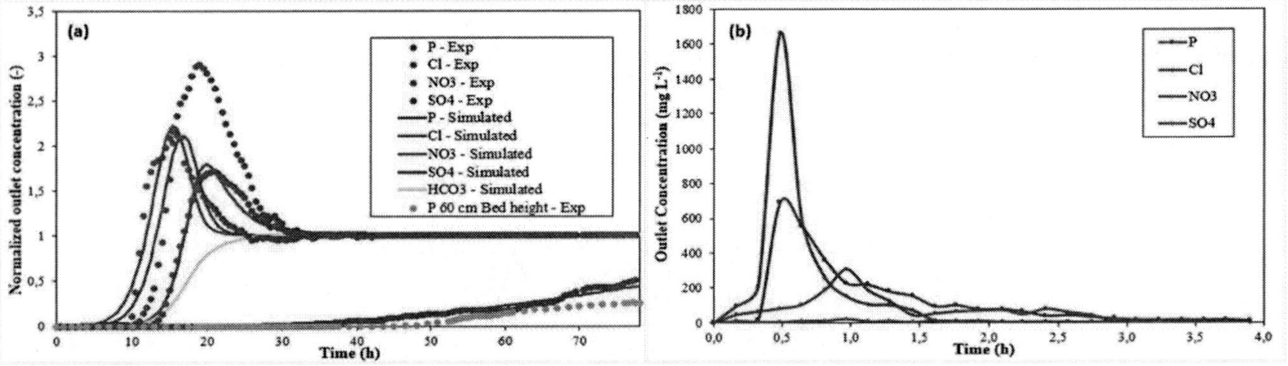

Figure 2. (a) P adsorption continuous flow test conducted with calcined pyroaurite in the lab-scale plant, pilot-plant (circles) and simulated (solid line). Experimental (circle) and simulated (solid line) breakthrough curves of phosphorous (blue - lab scale; orange- pilot-scale) and the competing anions: chloride (violet), nitrate (green), sulfate (red) and bicarbonate (yellow). (b) Experimental anions desorption breakthrough curves in the lab-scale plant. Breakthrough curves of phosphorous (blue), chloride (violet), nitrate (green) and sulfate (red).

Table 4. Operating conditions and performances of the breakthrough tests conducted with pyroaurite in the lab-scale and pilot plant.

	Test	Lab-scale	Pilot-plant
Operational Conditions	Bed height (m)	0.20	0.60
	EBCT, adsorption (min)	5.0	5.0
	EBCT, desorption (min)	10.0	10.0
	Regenerant type and concentration	NaOH 0.5M	NaOH 0.5M
Performances	BVs treated at BP 1 mg_P/L	641	715
	P adsorption yield 1 mg_P/L[a]	97%	98%
	Sorbent utilization efficiency at BP 1mg_P/L[b]	67%	55%
	P desorption yield (compared to adsorbed amount)[c]	69%	-
	P concentration in de desorbed product (mg/L)	201	-

a $\text{Adsorption yield} = \frac{\text{mass of P adsorbed at BP}}{\text{mass of P fed at BP}}$

b $\text{Sorbent utilization efficiency} = \frac{\text{mass of P adsorbed at BP}}{\text{mass of P fed at saturation}}$

c $\text{Desorption yield} = \frac{\text{mass of P adsorbed}}{\text{mass of P desorbed}}$

Table 5. Results of the parameter estimated with the Estimation function of Aspen Adsorption™.

	Value	Standard deviation
K_{AB} Cl (-)	2,03	0,142
K_{AB} HCO3 (-)	9,34	0,545
K_{AB} NO3 (-)	2,92	2,56
K_{AB} PO4 (-)	1,58	0,203
K_{AB} SO4 (-)	0,155	0,0208
$C_{L,HCO3}$ feed (eq/m³)	10,1	0,201
MTC (1/s)	0,07	-

402a Development of Solubility Kinetics for Removal of PFAS from Granular Activated Carbon with Supercritical Carbon Dioxide

Joseph T. Hilsendeger, Travis W. Walker, Kenneth M. Benjamin, Lori J. Groven

PFAS (per- and polyfluoroalkyl substances) compounds have contaminated many different matrices. Remediation of aquifers and other water sources have been one of the main focuses for research. Another major affected matrix exists in the solid form, such as solid waste, for example landfills and soil. These sources of contamination are both of major concerns because of the potential for PFAS compounds to leech into the undesirable areas due to the their resilient nature. Supercritical water oxidation (SWO) is a well-studied method for destruction of PFAS compounds. However, SWO of PFAS requires high temperatures and high pressures, and it is hazardous during operation because of the corrosive nature of SWO. Additionally, SWO may not selectively destroy PFAS when utilized in a crude feedstock, such as soil or hazardous waste.

A fluid is described as supercritical when the fluid is raised to a temperature and pressure that are above its critical temperature and pressure. When a fluid is supercritical, the fluid begins to have the diffusivity of a gas and the density of a liquid. Carbon dioxide is an abundant, low cost, and non-toxic fluid that has been used in supercritical fluid extraction (SFE) for decades. Carbon dioxide has a critical temperature of 31 degrees Celsius and a critical pressure of 74 bar. SFE with carbon dioxide has been shown to be effective in extracting nonpolar solids, and many PFAS compounds fit this profile.

SFE with carbon dioxide has been shown to be capable of removing a variety of PFAS compounds from various matrices. This extraction allows for a 'concentration' of solid waste as well as the remediation of the solid matrices that have been contaminated.

In this work, we experimentally develop solubility parameters for PFAS compounds in supercritical carbon dioxide at varying temperatures and pressures. These parameters are used to develop a theoretical model that can predict and screen for efficacy of SFE in removing PFAS from affecting matrices.

402b Optimization of Synthesis and Performance Evaluation of Natural Waste-Derived Adsorbent for PFAS Removal from Aqueous Solutions within a Closed-Loop Sustainable Process

Mohd Ahmed Naim Shaikh, Tabish Nawaz

Per- and poly-fluoroalkyl substances (PFASs) have been attracting worldwide attention in recent years due to their persistence nature, potential toxicity and high bioaccumulation. This study deals with adsorptive removal of PFAS from aqueous media. The activated carbon of waste Karanja (*Pongamia pinnata*) was optimally synthesized. Surface functional group modification was performed by using a cationic surfactant Cetyltrimethylammonium bromide (CTAB). Selectivity analysis of the adsorbent for 500 μg/L PFOA against different background species present at their typical river concentration was performed. To make the process sustainable, the development of regenerant specifications as well as recovery of the regenerant and PFOA from the spent regenerant solution were also investigated. A biomass/acid ratio (mg-Biomass per mL of 15M H_3PO_4) of 1:2.5, temperature 450 °C, and 1% CTAB solution was found to be optimum for the adsorbent synthesis for PFOA removal. The adsorption of PFOA was fast and equilibrium was achieved within 4 hours. The Langmuir model provided a good fit for the adsorption-isotherms data for both adsorbents. After functional group modification through CTAB, PFOA uptake capacity of the adsorbent increased up to threefold. The maximum PFOA adsorption capacity exhibited by Karanja activated carbon (K_AC) & CTAB-modified activated carbon (CTAB_K-AC) was 155.5 mg/g & 455.8 mg/g, respectively. The adsorbent exhibited high PFOA selectivity against all background co-contaminants i.e., NOM (humic acid), clay (bentonite), anions (Cl^-, NO_3^-, SO_4^{2-}, PO_4^{3-}, SiO_3^-, and CO_3^{2-}) and cations (Ca^{2+} and Mg^{2+}) and removed around 90% PFOA. The regenerant (50% ethanol with 1% NH_4OH aqueous solution) showed 80-90% PFOA recovery. The distillation process recovered 90% PFOA and 80-90% regenerant from the spent regenerant solution. The packed-bed column run of the adsorbent showed high PFOA removal capacity (>2000 bed volumes can be treated per gram of the adsorbent before break-through) from the river water matrix and the breakthrough data was well fitted with Thomas and Yoon Nelson Model with R^2 values of 0.97 for both the models. The results showed that CTAB_K-AC can be an effective adsorbent for PFOA adsorptive remediation from aqueous media with high treatment selectivity towards the PFOA in a materially closed-loop sustainable process.

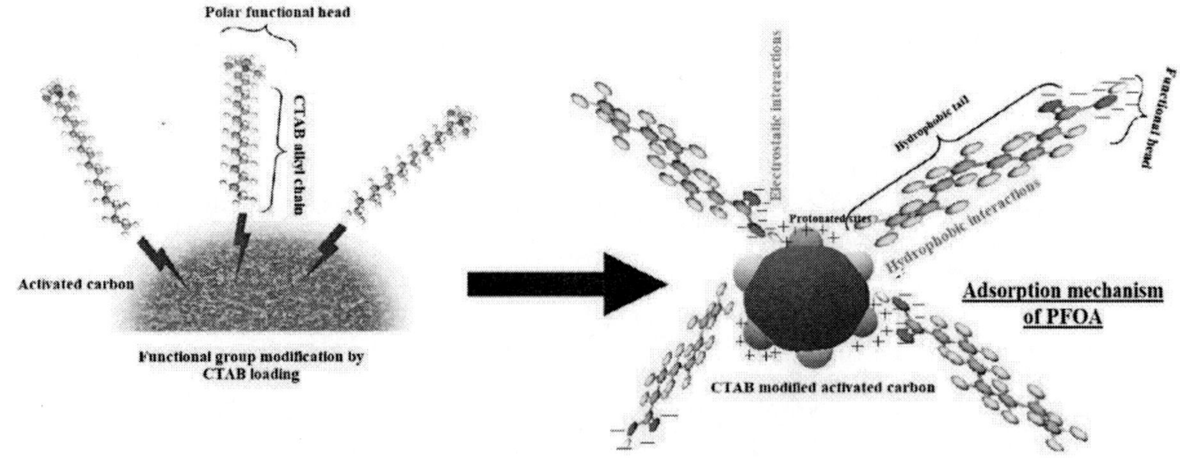

402c Biochar-Enabled Advanced Reduction Process for Enhanced Sorption and Degradation of per- and Polyfluoroalkyl Substances (PFAS) in Water

Dengjun Wang, Jianzhou He, Ziteng Song, Chongyang Wang, Nyla Metcalf, Kyra Anderson, Shahryar Jafarinejad

Per- and polyfluoroalkyl substances (PFAS) are manmade chemicals that are increasingly detected in the environment. PFAS exposure at low concentrations (e.g., ng/L) has been linked to various adverse effects on human health and the environment, so removing PFAS from the environment is urgently needed. However, due to the strong carbon-fluorine (C-F) bond in PFAS molecules, PFAS destruction is extremely challenging. The advanced reduction processes (ARPs) that generate reactive reducing radicals have been identified as an effective approach to breaking down the C-F bonds in PFAS by reductive reactions. We have been working on the development and optimization of ARPs for PFAS destruction and defluorination in water over the past years. We found that up to ~100% destruction efficiency and ~85% defluorination efficiency for different long-chain and short-chain PFAS in water can be achieved by the ARPs. The ARPs were sensitive to solution chemistry (e.g., pH, dissolved oxygen, and sulfite dosage) that controls the species and yield of reducing radicals responsible for PFAS destruction and defluorination. Adding biochars and modified biochars before ARPs was observed to enhance the destruction and defluorination efficiencies of PFAS in water due to the: (1) enhanced sorption (preconcentration) of PFAS on biochar surfaces; and (2) prolonged longevity for reducing radicals for extended reductive reactions between PFAS molecules and reducing radicals. These findings demonstrate that the ARPs, assisted by biochars, can be used as a viable strategy for the deep destruction and defluorination of various PFAS in water.

402d Use of Nitrogen-Doped Magnetic Quantum Carbon Dots for Photocatalytic Degradation of PFAS

Sanneri Santiago Borrés, Liu Yang, Ryan Gains, Katherine Deliz-Quinones, Andreia Fonseca de Faria

Exposure to PFAS has been linked to health outcomes including impaired function of kidney, liver, and thyroid, and suppression of immune, reproductive, and developmental systems. Since PFAS cannot be medically removed from the body, it is crucial to remove these toxic contaminants from the environment, to reduce risks on human health and ecosystem functions. PFAS contamination from wastewater discharge and reuse poses a significant threat to drinking water and agricultural water resources. Given the global distribution of PFAS, finding sustainable and cost-effective removal methods are essential. PFAS removal methods are still in development, and include both destructive and non-destructive techniques. Non-destructive approaches like activated carbon and reverse osmosis have shown limited success removing PFAS mixtures. Furthermore, they require frequent replacement of sorbents, leading to increased hazardous waste generation. Destructive techniques, like advanced oxidation and reduction processes (AOPs and ARPs) have higher success rates for removal of PFAS mixtures. However, they have high energy requirements, formation of potentially toxic byproducts, and scalability issues. Photocatalysis is a promising AOP that can be implemented at ambient temperature, with little energy requirements working even under sunlight, which increases sustainability and opportunities for scalability. We implemented a novel approach to cost-effectively develop urea-modified magnetic quantum carbon dots (M-QCD) from lignin waste, which exhibits high contaminant sorption. Preliminary findings show high quantum yield and production of reactive oxygen species, which improves the breakdown of PFAS mixtures from aqueous effluents. Our study will summarize the characterization and experimental results of M-QCD, highlighting its PFAS degradation potential.

402e Investigation of Advanced Oxidation Processes for Removal of Perfluorooctanoic Acid (PFOA) from Aqueous Matrices

Elizabeth Boyd, William E. Holmes, Rafael Hernandez, Daniel Gang, Mark E. Zappi

Perfluorooctanoic acid (PFOA) is a synthetic compound in the perfluoroalkyl substance (PFAS) family of chemicals. PFOA has been widely used in the production of Teflon, waterproof coatings, and many other industrial applications since the 1940s and has been identified as a human carcinogen by the United States Environmental Protection Agency (EPA). Agreements have been reached between major industrial users of this compound and the EPA to end the use of PFOA in the United States. This group of compounds is highly stable due to its chemical structure and does not easily biodegrade under natural conditions; thus, this persistent compound is found globally in ecosystems. The compound is also bioaccumulative. Current methods for removal of PFOA from water include reverse osmosis (RO) or adsorption, both of which concentrate the removed PFOA meaning that the concentrated material still needs to be handled and disposed of in a safe manner. Additionally, this removal is not particularly optimal. A majority of current research on removal of PFOA from aqueous matrices focuses on physical treatments including novel adsorbents.

Advanced oxidation processes (AOPs) are chemical oxidation processes that utilize hydroxyl radicals as the primary removal mechanism. They are classified as either dark or lighted AOPs. As the name suggests, dark AOPs do not utilize UV irradiation to catalyze formation of the hydroxyl radical from parent oxidizers while lighted AOPS utilize some form of UV irradiation to catalyze hydroxyl radical formation. Examples of dark AOPs include Fenton's Reagent and peroxone, and examples of lighted AOPs include low- and medium-pressure ultraviolet lamps used in conjunction with hydrogen peroxide (LPUV + HP and MPUV + HP). These processes have a long history of successful use in treatment of water contaminated with organic pollutants that are difficult to treat with other processes. While the chemical structure of PFOA makes oxidation of the compound potentially difficult, research on the oxidation of PFOA using optimized AOPs shows promise. One major benefit of AOPs for removal of organic pollutants over methods that concentrate the pollutant is its destruction of the pollutant.

A 3L bench-scale oxidation reactor was used for all experiments with the PFOA concentration set at 500 µg/L and the PFOA analyzed using high performance liquid chromatography with tandem mass spectroscopy (HPLC-MS/MS). The lighted AOPs investigated utilized a low pressure Hg-vapor lamp, a 200 watt medium pressure Hg-vapor lamp, and a 450 watt medium pressure Hg-vapor lamp. For each medium pressure lamp, both hydrogen peroxide concentration and reactor temperature were varied independently. A novel optimization scheme of hydrogen peroxide dosing was evaluated for optimizing radical production from the H_2O_2 and managing radical scavenging also from the H_2O_2. Additionally, experiments using ozone as the primary oxidizer in the system have been conducted to determine the efficacy of ozone in removing PFOA.

The results showed promise for removal of PFOA via MPUV + HP suggesting that AOPs could be effective for removal of PFOA in contaminated waters under proper conditions. Optimizing H_2O_2 dosing proved useful and a method process designers should consider for removing refractory pollutants and reducing operational costs.

402f Engineering High Frequency Ultrasound for Degradation of per- and Poly-Fluoroalkyl Substances

Yucheng Zhu, Mehrdad Zare, Auwal Yunusa, Madeleine Bussemaker

Per- and poly-fluorinated substances (PFAS) are man-made chemicals, used for more than 50 years in plastics, waterproofing, non-stick pans, surfactants and aqueous firefighting foams. Due to their widespread use, and extreme persistence PFAS are now everywhere in the environment. However, PFAS are toxic, bioaccumulate in animals and plants and can lead to numerous ill-health effects. Many PFAS are now banned or heavily regulated and hence legacy and environmental PFAS require destruction.

Many processes have been adapted for PFAS removal, such as nano-filtration, ozofractionation and adsorption. However, these generate concentrated PFAS streams which require further treatment, usually incineration at 1,000°C+ that emits HF and organofluorine compounds, which necessitate scrubbing [1]. PFAS contain a hydrophobic fluorinated carbon chain and hydrophilic head group(s) giving them desirable chemical properties which also makes them extremely recalcitrant to typical degradation methods. High frequency ultrasound mineralizes aqueous PFASs into less harmful and easily treatable species, such as CO_2 and F^- ions [2].

We will explore the challenges associated with sonolytic parameter selection using standard characterisation techniques. Presentation of sonoluminescence (SL) and sonochemiluminescence (SCL) to visualise and quantify the active bubble distribution will be presented alongside and iodide dosimetry to quantify the concentration of reactive oxidant species (ROS). A new image-processing-based quantitative method is introduced to overcome noise and glass reactor-reflected light interference in SL/SCL measurements. Typical ultrasonic behaviours will be discussed and results on use of machine learning to predict characterisation outcomes explored.

An overview of how sonolytic degradation of PFAS is expected to be useful toward total remediation of environmental contaminations will be presented. We focus on high frequency ultrasound (100 - 1,000 kHz) and examine parameters such as frequency, liquid height, power density and reactor configuration. Considerations of how high frequency ultrasound may be employed in an industrial setting will be discussed.

[1] I. Ross *et al.*, "A review of emerging technologies for remediation of PFASs," *Remediation*, vol. 28, no. 2, pp. 101–126, 2018.

[2] C. D. Vecitis, H. Park, J. Cheng, B. T. Mader, and M. R. Hoffmann, "Enhancement of perfluorooctanoate and perfluorooctanesulfonate activity at acoustic cavitation bubble interfaces," *J. Phys. Chem. C*, vol. 112, no. 43, pp. 16850–16857, 2008.

402g Control of Pore Sizes in Ordered Mesoporous Carbons for the Sorption of PFAS Chemicals

Pan Ni, Calvin Fales, Feng Xiao, Baolin Deng

Granular activated carbon (GAC) adsorption is commonly used for the removal of per-and polyfluoroalkyl substances (PFAS) from water. Its low sorption capacity and selectivity for short chain PFAS, e.g., perfluorobutanoic acid (PFBA), however, undermine its broad applications, especially, in the presence of natural organic matter (NOM) and other competing water constituents. This study aims to determine controlling mechanisms for PFAS sorption on carbon materials by developing ordered mesoporous carbons (OMCs) with precisely controlled pore sizes. Ordered microporous and mesoporous carbons with increasing pore sizes from 0.8, 1.6, 3.0, 4.2 to 5.3 nm were synthesized by a hard template method using Pluronic P-123 surfactant and tetraethyl orthosilicate. The performance testing results showed that the smaller the pore size, the better the removal performance toward short chain and long chain PFAS. For 0.8 nm OMCs, the sorption ability of perfluorooctanoic acid (PFOA) and PFBA increased by 102.9% and 378.7%, respectively, when compared to that of 5.3 nm OMCs. The observed sorption ability toward PFBA here is higher than most of the reported adsorbents, approaching 64 umol/g. When the pore size of an OMC is ≤ 3 nm, no significant inhibition was observed in both PFOA and PFBA sorption tests in the presence of 5 ppm NOM (humic acid).

717a Valorising Wastewater: A Novel Approach for Critical Raw Materials Recovery from Acid Mine Drainage

Tamlyn Naidu, Craig Sheridan, Peter Engelund Holm

Introduction:

Mining wastewater and Mining Influenced Water (MIW) represent formidable environmental challenges globally. Acidic mine drainage (AMD), in particular, has gained increased attention in recent years, following the 2015 Gold King Mine spill – leading to classification by the United Nations as the second largest environmental issue worldwide. This AMD and MIW concern extends beyond mining regions, impacting global resource management due to its pervasive and enduring legacy. Traditional AMD treatment methods, often chemical or energy-intensive such as reverse osmosis, are neither economically nor environmentally sustainable. Moreover, the prevalence of AMD intensifies climate change-related issues, including soil degradation, biodiversity loss, and adverse socio-economic effects. However, recent studies have revealed a (small) silver lining: many AMD streams contain substantial quantities of rare earth elements (REEs), and other critical raw materials (CRMs) such as arsenic. These waste streams also exhibit other characteristics that can be harnessed for value extraction, such as high acidity, rich iron content, and strategic geographic locations, often near other waste sources (proximity which presents opportunities for developing a circular economy by combining various waste streams). This study considers a perspective of broader AMD valorisation research, proposing an innovative approach to extract value from AMD, focusing on CRMs, particularly REEs. By integrating various treatment methodologies – ion exchange (IX) technology in particular for REE extraction – this approach redefines AMD from a costly waste liability to a valuable secondary resource. Such a paradigm shift not only mitigates the environmental impact of AMD but also potentially enhances global access to CRMs, addressing a critical need in the context of depleting natural resources.

Methods and Data:

This research focussed on the extraction of REEs from coal mine AMD in the Emalahleni region of South Africa (an area that has been impacted by both the social and environmental effects of MIW for decades). The research involved a detailed characterization of AMD to identify and quantify the presence of REEs. Ion-exchange (IX) technology was then evaluated as a method to extract these CRMs from the wastewater. The efficacy of different cationic and chelating resins, specifically chosen for their high affinity for trivalent cations, was assessed. Both batch and column studies were conducted to determine the effectiveness of these resins in extracting REEs from AMD. A thorough analysis of the adsorption capacities of different resins, an examination of adsorption isotherms and kinetics, and the assessment of desorption efficiencies was undertaken in this research.

Results:

AMD samples revealed concentrations of REEs with a recovery value of nearly 1 Euro/m^3 of AMD. It was found that both chelating and cationic resins closely followed the Langmuir adsorption isotherm model for most REE species, with the chelating resin also aligning to the Temkin model for Yttrium and Lanthanum. This indicated a strong interaction between adsorbate and adsorbent for REEs in AMD. The cationic resin displayed a greater overall adsorption efficiency (80 – 95%) than the chelating resin (33 – 45%) for the range of REEs tested – in the case of Gd, adsorbing up to 5 times more than the chelating resin. In batch kinetics studies, both resins exhibited pseudo-second order behaviour, suggesting that chemisorption is the rate-limiting step in both these systems. Findings from the breakthrough studies suggest that the cationic resin saturated with iron much sooner than with REEs – signifying that iron has a prominent effect on the performance. Desorption experiments with varying concentrations of EDTA and sulfuric acid demonstrated varying efficiencies, with higher concentrations of these eluates generally yielding better desorption results. Desorption was not entirely successful, especially with lower EDTA concentrations, indicating the need for further investigation.

Discussion and Take-Home Message:

This study underscores that AMD holds significant concentrations of valuable CRMs. IX technology proves effective for extracting these CRMs, with the research detailing the kinetic behaviour and efficiency of different IX systems. The presence of iron in AMD poses a challenge, affecting the efficiency of REE extraction and necessitating further research to enhance the process. Successful lab-scale results indicate potential for this method to address both wastewater treatment challenges and CRM recovery, promising transformative impacts if scaled effectively.

717b Treatment and Mineral Recovery from Combustion Residual Leachate and Produced Water

Chad Able, Thomas Schmitt, Alison Fritz

Critical minerals are key to a successful energy transition in the United States. Domestic sources of these minerals would increase energy security and enable more rapid decarbonization. Wastewater from fossil fuel sources contains varying amounts of critical minerals. This study analyzes two of these sources: leachate from coal combustion residual landfills and impoundments and produced water from unconventional sources. Treatment scenarios to facilitate mineral recovery via volume reduction were considered. Based on total dissolved solids, a membrane treatment system was analyzed for leachate, and a mechanical vapor compression system was considered for produced water. The Water Treatment Techno-economic Assessment Platform (WaterTAP) package in Python and the costing method developed by Process Optimization and Modeling for Minerals Sustainability (PrOMMiS) were used to analyze the cost and performance of both systems, whereas OLI Studio and Flowsheet enabled detailed electrolyte chemistry and brine analyses for both systems. The Combustion Product Information database published by the Electric Power Research Institute provided detailed constituent data for landfill and impoundment leachate, and the United States Geological Survey database on produced water provided by the National Energy Water Treatment Speciation database assisted with produced water analyses.
Treatment scenarios considered in this work calculated a range of levelized cost of water values (based on the flow rate of clean water) between \$1.85 and \$15.45/m^3 for combustion residual leachate and between \$14.20 and \$24.10/m^3 for produced water in 2023 dollars. The value of the resultant brine was between \$1.64 and \$15.99/m^3 for leachate and between \$6.52 and \$34.87/m^3 for produced water. In both cases, this value was primarily dependent on the value of lithium and magnesium in the feed. Based on these values and the resultant recovery percentages (95% for leachate and 50% for produced water from the Permian Basin), treatment for the sake of recovery is suggested only for produced water. Magnesium recovery from water is already performed industrially for seawater and other brines. Direct lithium extraction will be considered in future work.

717c Implications of Physicochemical Structure on Efficient Valorization of Candlenut Shells

Heather LeClerc, Esther D. Mwakisambwe, Hanno C. Erythropel, Paul T. Anastas, Julie B. Zimmerman

Opponents to biomass valorization cite the "food vs. fuel" conflict[1] to avoid using the 140 Gigatons of wasted biomass generated around the world annually.[2] The utilization of the non-edible fraction of biomass waste therefore represents an abundant, renewable feedstock to address the growing fuel and chemical demands of society. Utilization of these non-edible biomass wastes has the potential to produce value-added products while simultaneously diverting waste from landfills as well as its subsequent decomposition emissions.

Biomass is primarily composed of cellulose, hemicellulose, and lignin, wherein lignin serves as the largest natural sources of aromatic compounds on Earth, typically accounting for 15-30 wt% of plant matter.[3] Nut shells, on the other hand, have reported up to 40% lignin[4] content. In 2012, vanilla seed coats and other nut shells were found to contain a novel homopolymer of caffeyl alcohol, coined C-lignin.[5] C-lignin is unique as it is often found as a homopolymer bound by benzodioxane linkages and potentially depolymerizes to the single product catechol[5] which is a useful precursor molecule for proteins[6] and biomimetic materials.[7] Candlenut shells (CNS) from the Euphorbiaceae family are one such source of C-lignin that further represents a centralized feedstock due to current processing of the internal seed. Compared to traditional woody biomass, the physiological role and physicochemical properties of the candlenut shell introduce significant challenges to its effective valorization.

This work aims to understand the role of CNS physicochemical structure on its valorization potential through a combined biological and chemical lens. Candlenut shells are comprised of specialized macrosclereid cells containing a lignified secondary cell wall and designed to restrict water uptake and protect the seed. This cell type (**Figure 1a**) gives CNS a distinct non-porous structure that first, significantly differs from traditional biomass, and second, severely inhibits effective extraction. Due to its nonporous structure, lignin extraction has been found to directly correlate with particle surface area. By increasing surface area from 0.36 m^2/g to 1.32 m^2/g, lignin extraction yield increases 98%. Furthermore, increased extraction time indicates an asymptote (**Figure 1b**) wherein increasing the solvent-biomass ratio does not result in additional lignin yield. Relating this finding back to CNS structure reveals the presence of cutin and suberin moieties that effectively block further lignin recovery due to the lack of internal pores. This work utilizes multiple organic solvents in sequential extraction steps along with a range of characterization techniques to uncover the role of candlenut shell physicochemical structure on their efficient valorization through a quantitative analysis of both lignin and suberin/cutin extraction components and the effect of solvent properties.

References:

1. Abdul Hakim Shaah, M.; Hossain, M. S.; Salem Allafi, F. A.; Alsaedi, A.; Ismail, N.; Ab Kadir, M. O.; Ahmad, M. I., A review on non-edible oil as a potential feedstock for biodiesel: physicochemical properties and production technologies. *RSC Adv* **2021,** *11* (40), 25018-25037.
2. Tripathi, N.; Hills, C. D.; Singh, R. S.; Atkinson, C. J., Biomass waste utilisation in low-carbon products: harnessing a major potential resource. *npj Climate and Atmospheric Science* **2019,** *2* (1), 35.
3. Zakzeski, J.; Bruijnincx, P. C. A.; Jongerius, A. L.; Weckhuysen, B. M., The Catalytic Valorization of Lignin for the Production of Renewable Chemicals. *Chemical Reviews* **2010,** *110* (6), 3552-3599.
4. Queirós, C. S. G. P.; Cardoso, S.; Lourenço, A.; Ferreira, J.; Miranda, I.; Lourenço, M. J. V.; Pereira, H., Characterization of walnut, almond, and pine nut shells regarding chemical composition and extract composition. *Biomass Conversion and Biorefinery* **2020,** *10* (1), 175-188.
5. Li, Y.; Shuai, L.; Kim, H.; Motagamwala, A. H.; Mobley, J. K.; Yue, F.; Tobimatsu, Y.; Havkin-

Frenkel, D.; Chen, F.; Dixon, R. A.; Luterbacher, J. S.; Dumesic, J. A.; Ralph, J., An "ideal lignin" facilitates full biomass utilization. *Science Advances 4* (9), eaau2968.

6. Yang, J.; Cohen Stuart, M. A.; Kamperman, M., Jack of all trades: versatile catechol crosslinking mechanisms. *Chemical Society Reviews* **2014,** *43* (24), 8271-8298.
7. Sedó, J.; Saiz-Poseu, J.; Busqué, F.; Ruiz-Molina, D., Catechol-Based Biomimetic Functional Materials. *Advanced Materials* **2013,** *25* (5), 653-701.

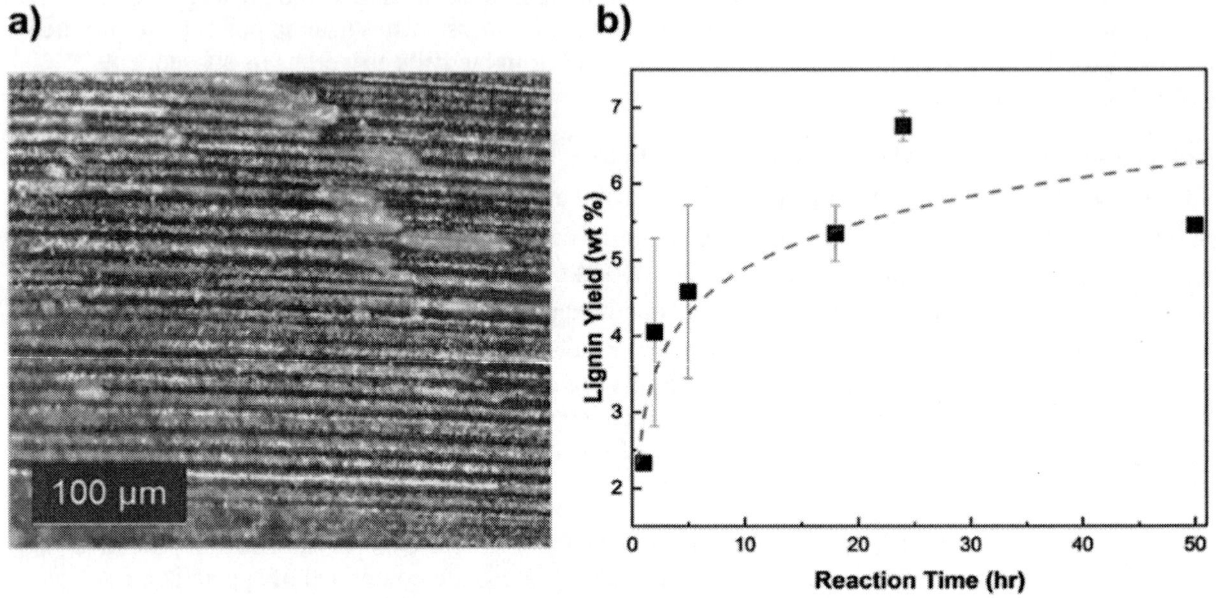

403a Hydrothermal Liquefaction of Plastic Waste

Rajarshi Bandyopadhyay, Kalsoom Jan, Akash Patil, Rodrigo Almeida Luciano, Daniel Lachos Perez, Grace Chen, Débora M. Vaz de Miranda, Andressa Improta Cruz

The escalating production and disposal of plastic waste pose a significant environmental challenge in recent times. Conventional waste management strategies have often neglected plastic waste's potential as a renewable energy source and petrochemical feedstock. Hydrothermal liquefaction (HTL), a promising technology, has emerged as a sustainable solution to address these challenges. HTL is distinguished by its use of water as a solvent, leveraging environmental and economic benefits in transforming waste feedstock into an oil product (HTL oil), that is similar to the compositional characteristics of crude petroleum. The aim of the present study is the application of HTL for converting polyethylene (PE) waste into HTL oil, thereby presenting a novel approach to managing plastic waste.

Polyethylene (PE) was chosen because it is one of the most common consumer plastics and forms a major fraction of municipal plastic waste. In the present topic, polyethylene pellets were utilized as a feedstock in HTL processes to produce oil products, investigating the impact of various operating conditions on oil yield. Plastic waste is often contaminated, these impurities can be inherent in the form of additives used to render different physical characteristics to the consumer product. The presence of such additives/impurities can compromise the recycling efforts. To combat the issue, experimental runs were also carried out with samples of PE mixed with a known impurity, and the impact of such impurities was then monitored on oil yield (and composition) to determine the best HTL protocol.

Our findings reveal that HTL oil with a composition and purity suitable for use in fluidized catalytic crackers can be produced, effectively reducing the net demand for crude oil in refineries. This study underscores the potential of HTL in transforming plastic waste into valuable HTL oil, contributing to a circular economy, and offering a sustainable alternative to conventional waste management and energy production methods.

403b Molecular Recycling of Multilayer Plastic Films Via Delamination

Jacob Licht, Christian Ferger, Marina Tsianou, Paschalis Alexandridis

40% of all plastic is used in packaging, with a major fraction of this in the form of films. Most plastic films comprise multiple layers that impart desirable functionality (e.g., oxygen or moisture barriers, labels). Multilayer films are not mechanically recycled. Multilayer films can be processed via pyrolysis, the most common among chemical recycling processes, this, however, involves the break-down of polymers into fuel or chemical feedstock, resulting in undesirable greenhouse gas (GHG) emissions. Plastics recycling research in our group utilizes physical, solvent-based processes whereby polymer chains do not break. This constitutes true recycling, as the recovered polymer is the same as the staring material.

The objective of this project is to develop solvent-based processes to deconstruct flexible multilayer plastic films while leaving intact (in its solid form) the main component polyolefins. Specifically, we develop strategies based on delamination for separating individual polymers from multilayer packaging materials, we recover solid (undissolved) polyolefins via gravity separation, and perform a preliminary techno-economic analysis of the whole process.

The solvent-based multilayer film delamination process developed in this project is environmentally responsible on the basis of low energy usage and GHG emissions. The recovered polyolefins, following appropriate processing, can replace primary materials without loss of properties or performance and, hence, meet demand by customers and corporations to incorporate recycled plastics into products.

403c From Degradation Kinetics to Process Evaluation: Challenges and Solution Approaches in Designing Chemical Recycling Processes

Philip Biessey, Meike Holtkamp, Darius Steegborn

The chemical industry is more and more confronted with social and political expectations to increase process efficiency and to provide new process technologies, allowing for the utilization and implementation of previously unused material streams such as plastic waste. For that purpose, circular economy principles in terms of circularity strategies are considered increasingly: besides the utilization of circularity strategies such as product re-design (which can be driven e.g. by means of simplified recycling approaches) and the avoidance and/ or substitution of crude oil derivatives and other fossil resources, the focus is strongly on recycling processes. In the chemical industry, such processes aim at the reintegration of material flows into specific production processes, e.g. the integration of pyrolysis oils in steam crackers as a naphtha substitute [1] or the re-integration of monomers, obtained from chemical recycling processes, in polymer production [2, 3]. Regarding plastics in particular, chemical recycling technologies are currently strongly focussed: The complementary reintegration of material flows from both mechanical and chemical recycling processes aims at substituting crude oil derivatives and allowing for closed material cycles, thus reducing waste volumes and fossil resource depletion in the genuine sense of a circular economy.

To date, there are three main limitations and challenges in the design and engineering of chemical recycling processes: first, there is a lack of systematic methodologies for the implementation and combination of both experimental data - especially regarding kinetics of the conversion steps - and model-based approaches for the design and scaling of reactors and entire processes for chemical recycling technologies (challenge #1). This is a crucial limitation since the availability of a design tool for reactors and entire processes is crucial for knowledge-based engineering. Secondly, the varying composition of the chemical recycling products such as pyrolysis oils, which is a consequence of different waste sources and process conditions, requires robust downstream processes to ensure successful reintegration of the obtained products and material streams. For both process robustness, and chemical recycling downstream design, no knowledge-based approaches are available (challenge #2). And finally, an environmental assessment of possible reintegration pathways is essential to assess the sustainability at the earliest possible stage of process development; this aspect is challenging because it must typically be carried out at a very early stage of process development, in which data availability and data quality is often not given (challenge #3). **Based on polystyrene as an example, this contribution presents different solution approaches to overcome the above-mentioned challenges in the design and evaluation of chemical recycling processes.**

To define and identify suitable operating windows of processes, and to implement experimental data especially obtained from plastic waste degradation investigations, process simulation tools using computer-based modelling approaches are often employed. Based on validated process simulations, these processes can be scaled, optimized, and evaluated regarding, e.g., their energy consumption and environmental impacts without the need for costly and time-consuming experimental trials. In chemical recycling processes, the knowledge-based definition and identification of operating windows considering different and varying types of feedstocks is of particular importance. Process simulation can help identifying such operating windows by using computer-based modelling to understand and predict the behavior of both conversion steps and downstream operations. But simulating especially the reaction networks of chemical recycling processes is challenging due to the complexity of the reaction system. For example, pyrolysis processes involve a complex reaction network, resulting in the formation of various reactive intermediates, that lead to a large variety in the final product composition. The sheer number of possible reaction pathways makes it difficult, if not impossible, to solve the stoichiometry of the entire reaction system. Consequently, the reaction systems must be treated as black boxes in which the feedstock reacts to the actual products. To achieve a knowledge-based process simulation, the reaction model must contain kinetic data representing the actual behavior of the reaction system. The derivation of those kinetic data sets for the degradation of polymers is addressed in numerous studies ([4–6], among others), mostly using pure polymer fractions or defined polymer mixtures, some already gathering kinetic data from actual waste fractions. Those waste fractions, especially post-consumer wastes, are not homogenous. The composition, the particle sizes and the grade of impurity differ a lot through the year influenced by the season, the collection area, and the sorting facilities. Therefore, further investigation on how these varying feed-parameters affect the reaction is needed, which is addressed in this contribution: Kinetic data sets are derived and compared for real waste polystyrene fractions collected from sorted light packaging waste using simultaneous thermal analysis measurements, providing a solution approach for challenge #1. Two polystyrene samples are taken before and after a cleaning process resulting in two fractions with a defined grade of impurity. Each is sorted by particle size to investigate the influence of the impurities on the degradation and product yield of the reaction. The results are used to overcome the stated limitations giving suitable kinetic data sets for process simulations and development.

The resulting varying composition of the pyrolysis oil due to the different waste composition requires robust downstream processes to ensure successful product reintegration. Thus, to provide a solution approach for challenge #2, different pyrolysis oil compositions obtained from waste streams with varying styrene content are investigated in a downstream process simulation using sensitivity analysis. In the Aspen Plus® based process simulation, process robustness is evaluated in terms of the sensitivity of overall energy demand and product quality towards distillation process parameters: reflux ratio, distillate-to-feed ratio, and side-stream control. Depending on the pyrolysis oil composition two different reintegration pathways for the main product styrene as well as the by-products are considered in the distillation sequence.

Finally, a prospective life-cycle assessment (LCA) is conducted comparing the chemical recycling pathways to waste incineration at industrial scale. The chemical recycling route is shown to be favorable over the waste incineration considering the impact categories global warming potential (GWP) and fossil fuel depletion potential (FFDP). The prospective LCA also identifies environmental trade-offs by including the impact categories freshwater eutrophication (FE), marine eutrophication (ME) and terrestrial acidification (TE). The FE and ME potential is much lower for waste incineration, although this is mainly due to the high eutrophication potential of the current electricity mix. Finally, the prospective LCA results of the possible reintegration pathways are presented to assess the sustainability at the earliest possible stage of process development and discuss methodological limitations and possibilities based on this example as a solution approach for challenge #3.

In summary, this contribution gives detailed insights into major challenges and solution approaches for the design of chemical recycling processes, mainly focusing on the integration of experimentally obtained reaction kinetics, model-based process development and prospective LCA. Based on that, the applicability of the proposed methods and approaches are evaluated with respect to the forthcoming shift towards a circular plastics economy.

[1] M. Kusenberg, G. C. Faussone, H. D. Thi, M. Roosen, M. Grilc, A. Eschenbacher, S. de Meester, K. M. van Geem, *The Science of the total environment* **2022**, *838 (Pt 2)*, 156092. DOI: https://doi.org/10.1016/j.scitotenv.2022.156092

[2] A. Merkel, L. Plessing, G. A. Luinstra, M. Grünewald, P. Biessey, *Chemie Ingenieur Technik* **2023**, *95 (8)*, 1323 – 1331. DOI: https://doi.org/10.1002/cite.202100210

[3] M. Holtkamp, M. Renner, K. Matthiesen, M. Wald, G. A. Luinstra, P. Biessey, *Resources, Conservation and Recycling* **2024**, *205*, 107558. DOI: https://doi.org/10.1016/j.resconrec.2024.107558

[4] W. Zhang, J. Zhang, Y. Ding, Q. He, K. Lu, H. Chen, *J. Cleaner Prod.* **2021**, *285*, 125042. DOI: https://doi.org/10.1016/j.jclepro.2020.125042

[5] J. Nisar, G. Ali, A. Shah, M. Iqbal, R. A. Khan, Sirajuddin, F. Anwar, R. Ullah, M. S. Akhter, *Waste management (New York, N.Y.)* **2019**, *88*, 236 – 247. DOI: https://doi.org/10.1016/j.wasman.2019.03.035

[6] X. Ren, Z. Huang, X.-J. Wang, G. Guo, *J Therm Anal Calorim* **2022**, *147 (2)*, 1421 – 1437. DOI: https://doi.org/10.1007/s10973-020-10411-9

403d Enhancing Polystyrene Upcycling through Sbsi Chalcohalides-Driven Photocatalysis: A Novel Approach to Combat Plastic Waste

Goutham Rangarajan, Rohit Pal, Ramin Farnood

Plastic waste management remains a critical environmental challenge. Since 1950, over 6.3 billion tons of plastic have been produced, with more than 80% ending up in landfills or polluting the natural environment [1]. Polystyrene (PS) consumer products worsen this issue due to their extensive applications and remarkably short lifecycle. Given the rising demand for these materials and their difficulty to be recycled, there is an interest in employing photocatalysis as a sustainable method to depolymerize PS into valuable products [2]. Nevertheless, the application of conventional semiconductor photocatalysts in PS upcycling has been limited due to wide bandgaps and charge recombination.

Chalcohalides, a novel family of inorganic semiconductors, emerges as a promising solution. These materials feature highly dispersive band edges that facilitate efficient charge transfer and exhibit narrow bandgaps [3]. Also, the presence of divalent chalcogen and monovalent halide anions in these materials provides enhanced defect tolerance.

This study demonstrates that SbSI chalcohalide can generate singlet oxygen ($^1O_2$), a potent, non-radical reactive oxygen species. $^1O_2$ possesses a high energy (~94 kJ/mol) compared to ground-state O_2, enabling it to abstract hydrogen from the stable benzylic C(sp^3)–H bond in PS [4]. This initiates degradation of PS into valuable oxygenated products such as benzaldehyde and acetophenone under mild reaction conditions. Preliminary experiments with PS dimer model compound (1,3-diphenylpropane) showed 28.37% C_7-product yield with ~80% selectivity towards benzaldehyde. Detailed experimental results focusing on the reaction mechanism, supported by catalyst characterization, Electron Paramagnetic Resonance (EPR) spectroscopy, and untargeted Ultra High Performance Liquid Chromatography-Tandem Mass Spectrometry (UHPLC-MS/MS) analysis will be presented at the conference.

References

[1] Alabi, O. A., Ologbonjaye, K. I., Awosolu, O., & Alalade, O. E. (2019). Public and environmental health effects of plastic wastes disposal: a review. *J Toxicol Risk Assess, 5*(021), 1-13.

[2] Ghalta, R., Bal, R., & Srivastava, R. (2023). Metal-free photocatalytic transformation of waste polystyrene into valuable chemicals: advancing sustainability through circular economy. *Green Chemistry, 25*(18), 7318-7334.

[3] Ran, Z., Wang, X., Li, Y., Yang, D., Zhao, X. G., Biswas, K., ... & Zhang, L. (2018). Bismuth and antimony-based oxyhalides and chalcohalides as potential optoelectronic materials. *npj Computational Materials, 4*(1), 14.

[4] Huang, Z., Shanmugam, M., Liu, Z., Brookfield, A., Bennett, E. L., Guan, R., ... & Xiao, J. (2022). Chemical recycling of polystyrene to valuable chemicals via selective acid-catalyzed aerobic oxidation under visible light. *Journal of the American Chemical Society, 144*(14), 6532-6542.

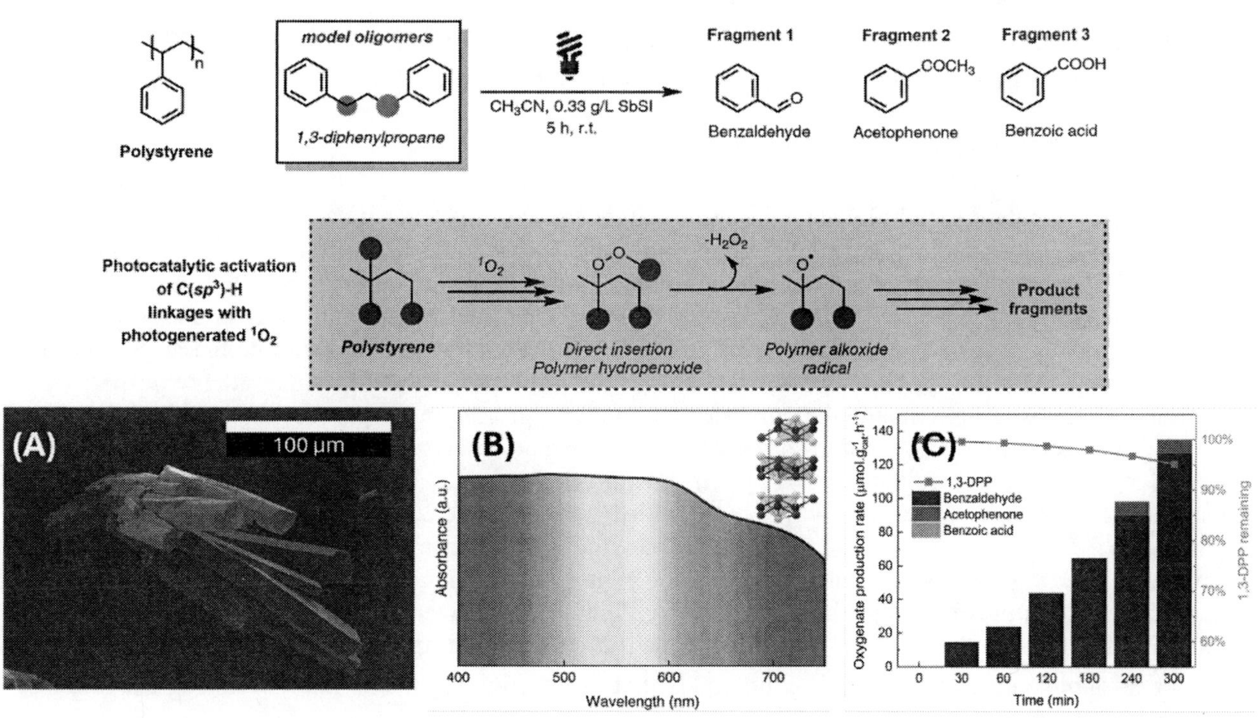

Schematic representation of the photocatalytic conversion of polystyrene and its dimer model compound (1,3-diphenylpropane) with SbSI photocatalyst induced singlet oxygen. (A) SEM image of SbSI particles, (B) Visible-light absorption capacity of SbSI nanoparticles, (C) Representative conversion of 1,3-DPP into oxygenates under O_2 and light.

403e Embedding an Esterase Mimic inside Polyesters to Realize Rapid and Complete Degradation without Compromising Their Utility

Qikun Hu, Yanfen Wu, Zhiqiang Niu

In the face of the escalating global plastic pollution, considerable efforts have been devoted to the development of biodegradable plastics over the past decades, aiming to substitute persistent commodity plastics. However, the degradation rate of biodegradable plastics is remarkably slow in the environment. Developing biodegradable plastics with accelerated degradation under both composting and ambient conditions is important for the mass use of such materials. Current strategies to promote their degradation often compromise their utility. Here, we enhance the degradation properties of biodegradable plastics by embedding an esterase mimic inside the polyester without compromising its processing and mechanical properties. This esterase mimic contains a binuclear zinc site and catalyzes the depolymerization of poly(butylene terephthalate/adipate) (PBAT) with a turnover frequency (TOF) an order of magnitude higher than that of conventional additives (ZnO). With 1 wt% of this enzyme mimic embedded, PBAT exhibits a 3.7-fold enhancement of the degradation rate under composting conditions. Mechanistic investigation indicates that the binuclear zinc binding center polarizes the ester carbonyl group and increases hydroxide uptake, thus leading to faster degradation. Moreover, cytotoxicity studies confirmed that the PBAT/Zn_2L composite maintains an equivalent level of biocompatibility to pure PBAT. This work provides a possible approach to designing polymers that meet requirements of stability during use and fast degradation after use.

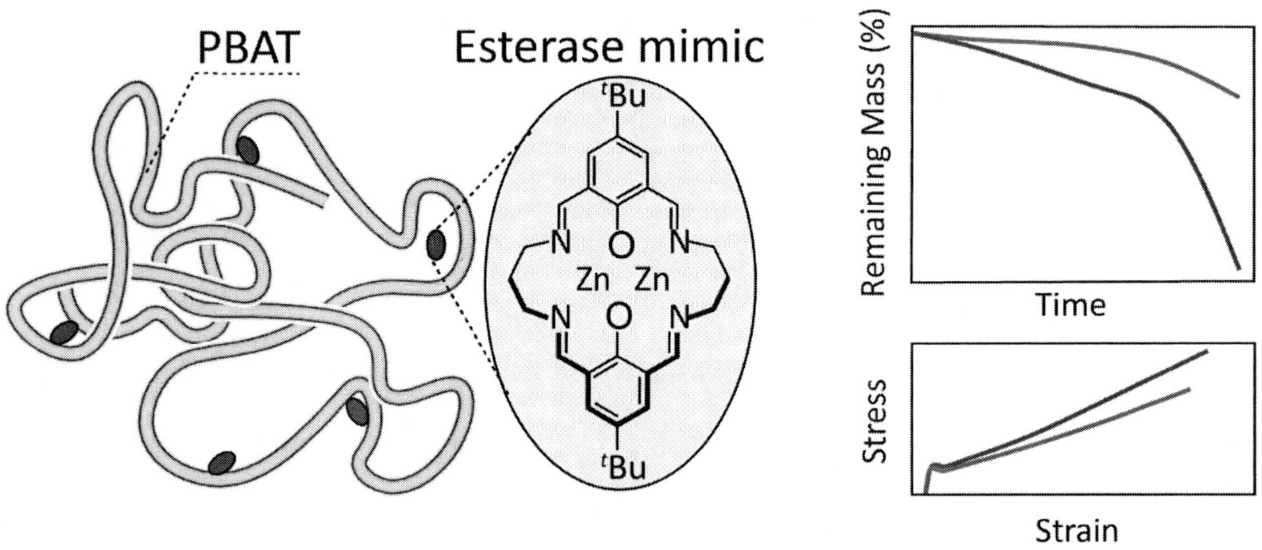

403f Agile Techno-Economic Analysis to Advance Circular Manufacturing of Poly-Crude from Waste Plastics through Catalytic Hydrocracking

Elizabeth Aigaje, Rui Shi

There is a clear call for the advancement in innovative solutions that add value to plastic waste. Chemical recycling has gain attention for its ability to transform plastic waste into fuels and chemical compounds, which has the potential to not only helps to reduce the carbon footprint and energy consumption by valorizing plastic waste into value-added products, but also aligns with efforts to achieve a circular plastics economy. Thermal chemical conversion of plastics refers to the process of converting plastic waste into energy or useful chemicals through thermal treatment methods such as pyrolysis, hydrothermal liquefaction (HTL), solvolysis, hydrogenolysis, hydrocracking, etc. [1]. Among these, hydrocracking has the advantages of low operation temperatures, high production of alkanes over olefins, and low coke formation. Moreover, hydrocracking yields high-quality liquid oil without the need of several separation steps [2]. Additionally, it is a well-established process in the refinery industry that alleviates the challenges in scale-up and infrastructure, and therefore has recently gained attention for its efficacy in decomposing plastics, particularly for polyolefins [3], [4], [5].

Despite the continuous progress in catalysts design and reaction performance, a substantial gap persists in comprehending how factors such as processing design, operating conditions, uncertainties in materials price, and others technical and economic variables may impact on the economic viability of the process. In this regard, this research performs system level techno-economic analysis (TEA) and life cycle assessment (LCA) to elucidate the environmental impacts and economic potential of the potential manufacture of poly-crude, a compatible product with the current refinery feedstock, from waste POs. We aim to conduct agile TEA at the early stage of the technology development to provide guidance in identifying the operational range where plastic conversion through hydrocracking is economically feasible. The integration with current Refinery can take advantage of the already existing petrochemical facilities which represents enormous savings on capital investment [6]. We evaluate new hydrocracking process economics in the context of the catalytic pathways that emerge from recent studies, following established approaches for cost analysis using n^{th}-plant project costing and financing [7], [8]. Discounted cash flow analysis are used to estimate final product pricing. Preliminary TEA based on literature inputs and market reports indicate process economics are strongly impacted by on-site H_2 storage and PO feedstock costs (note feedstock costs apply to other chemical conversion processes as well). Potentially favorable poly-crude process economics were identified for feedstock prices near $500/ton for 250-500 ton/day plant capacities. Other decisions and process parameters such as feed conversion and liquid yield, also influence the economic metrics. Preliminary results indicate that feedstock price may account for approximately 60-70% of the MCSP. For this study we plan to establish connections between catalyst performance metrics (e.g., yield, conversion) and key economic indicators such as minimum crude selling price (MCSP), net present value (NPV), internal rate of return (IRR), among others, which will allow researchers to identify critical process parameters and product considerations governing the economic feasibility of integrating waste POs into the petrochemical supply chain

References

[1] O.D. Bozkurt, S. T. Okonsky, K. Alexopoulos, and H. E. Toraman, 'Catalytic conversion of SPW and products upgrading', *in Advances in Chemical Engineering*, vol. 60, Elsevier, 2022, pp. 117-168. doi: 10.1016/bs.ache.2022.09.006.

[2] D. Munir, M. F. Irfan, and M. R. Usman, 'Hydrocracking of virgin and waste plastics: A detailed review', *Renewable and Sustainable Energy Reviews*, vol. 90, pp.490-515, Jul. 2018, doi: 10.1016/j.rser.2018.03.034.

[3] L. Li *et al.*, "Converting Plastic Wastes to Naphtha for Closing the Plastic Loop," *J. Am. Chem. Soc.*, vol. 145, no. 3, pp. 1847–1854, Jan. 2023, doi: 10.1021/jacs.2c11407.

[4] S. Liu, P. A. Kots, B. C. Vance, A. Danielson, and D. G. Vlachos, "Plastic waste to fuels by hydrocracking at mild conditions," *Sci. Adv.*, vol. 7, no. 17, p. eabf8283, Apr. 2021, doi: 10.1126/sciadv.abf8283.

[5] Z. Qiu *et al.*, "A reusable, impurity-tolerant and noble metal–free catalyst for hydrocracking of waste polyolefins," *Sci. Adv.*, vol. 9, no. 25, p. eadg5332, Jun. 2023, doi: 10.1126/sciadv.adg5332.

[6] G. Yadav *et al.*, "Techno-economic analysis and life cycle assessment for catalytic fast pyrolysis of mixed plastic waste," *Energy Environ. Sci.*, vol. 16, no. 9, pp. 3638–3653, 2023, doi: 10.1039/D3EE00749A.

[7] Davis, R. et al. 'Process Design and Economics for the Conversion of Lignocellulosic Biomass to Hydrocarbons: Dilute-Acid and Enzymatic Deconstruction of Biomass to Sugars and Catalytic Conversion of Sugars to Hydrocarbons'. Report No. NREL/TP-5100-62498 United States 10.2172/1176746 NREL English, Medium: ED; Size: 133 pp. (National Renewable Energy Laboratory (NREL), Golden, CO., 2015,

[8] Phillips, S. D., Tarud, J. K., Biddy, M. J. & Dutta, A. 'Gasoline from Woody Biomass via Thermochemical Gasification, Methanol Synthesis, and Methanol-to-Gasoline Technologies: A Technoeconomic Analysis'. I*nd. Eng. Chem*. Res. 50, pp.11734-11745, 2011.

AUTHOR INDEX

AIChE
120 Wall Street, FL 23
New York, NY 10005-4020

ISBN 979-8-3313-1678-5

CURRAN ASSOCIATES INC.
proceedings.com